OPTIMAL SPORTS MATH, STATISTICS, AND FANTASY

OPTIMAL SPORTS MATH, STATISTICS, AND FANTASY

ROBERT KISSELL

AND

JIM POSERINA

ACADEMIC PRESS

An imprint of Elsevier
elsevier.com

Academic Press is an imprint of Elsevier
125 London Wall, London EC2Y 5AS, United Kingdom
525 B Street, Suite 1800, San Diego, CA 92101-4495, United States
50 Hampshire Street, 5th Floor, Cambridge, MA 02139, United States
The Boulevard, Langford Lane, Kidlington, Oxford OX5 1GB, United Kingdom

Notice

No responsibility is assumed by the publisher for any injury and/or damage to persons or
property as a matter of products liability, negligence or otherwise, or from any use or operation
of any methods, products, instructions or ideas contained in the material herein. Because of
rapid advances in the medical sciences, in particular, independent verification of diagnoses and
drug dosages should be made.

British Library Cataloguing-in-Publication Data
A catalogue record for this book is available from the British Library

Library of Congress Cataloging-in-Publication Data
A catalog record for this book is available from the Library of Congress

ISBN: 978-0-12-805163-4

For information on all Academic Press publications
visit our website at https://www.elsevier.com/books-and-journals

Working together
to grow libraries in
developing countries

www.elsevier.com • www.bookaid.org

Publisher: Nikki Levy
Acquisition Editor: Glyn Jones
Editorial Project Manager: Lindsay Lawrence
Production Project Manager: Omer Mukthar
Cover Designer: Mark Rogers

Typeset by MPS Limited, Chennai, India

CONTENTS

BIOGRAPHICAL INFORMATION

Dr. Robert Kissell is the President and founder of Kissell Research Group. He has over 20 years of experience specializing in economics, finance, math & statistics, algorithmic trading, risk management, and sports modeling.

Dr. Kissell is author of the leading industry books, "The Science of Algorithmic Trading & Portfolio Management," (Elsevier, 2013), "Multi-Asset Risk Modeling" (Elsevier, 2014), and "Optimal Trading Strategies," (AMACOM, 2003). He has published numerous research papers on trading, electronic algorithms, risk management, and best execution. His paper, "Dynamic Pre-Trade Models: Beyond the Black Box," (2011) won Institutional Investor's prestigious paper of the year award.

Dr. Kissell is an adjunct faculty member of the Gabelli School of Business at Fordham University and is an associate editor of the *Journal of Trading* and the *Journal of Index Investing*. He has previously been an instructor at Cornell University in their graduate Financial Engineering program.

Dr. Kissell has worked with numerous Investment Banks throughout his career, including UBS Securities where he was Executive Director of Execution Strategies and Portfolio Analysis, and at JPMorgan where he was Executive Director and Head of Quantitative Trading Strategies. He was previously at Citigroup/Smith Barney where he was Vice President of Quantitative Research, and at Instinet where he was Director of Trading Research. He began his career as an Economic Consultant at R.J. Rudden Associates specializing in energy, pricing, risk, and optimization.

During his college years, Dr. Kissell was a member of the Stony Brook Soccer Team and was Co-Captain in his Junior and Senior years. It was during this time as a student athlete where he began applying math and statistics to sports modeling problems. Many of the techniques discussed in "Optimal Sports Math, Statistics, and Fantasy" were developed during his time at Stony Brook, and advanced thereafter. Thus, making this book the byproduct of decades of successful research.

Dr. Kissell has a PhD in Economics from Fordham University, an MS in Applied Mathematics from Hofstra University, an MS in Business Management from Stony Brook University, and a BS in Applied Mathematics & Statistics from Stony Brook University.

Dr. Kissell can be contacted at info@KissellResearch.com.

R. Kissell

Jim Poserina is a web application developer for the School of Arts and Sciences at Rutgers, the State University of New Jersey. He has been a web and database developer for over 15 years, having previously worked and consulted for companies including AT&T, Samsung Electronics, Barnes & Noble, IRA Financial Group, and First Investors. He is also a partner in Doctrino Systems, where in addition to his web and database development he is a systems administrator.

Mr. Poserina has been a member of the Society for American Baseball Research since 2000 and has been published in the *Baseball Research Journal*. He covered Major League Baseball, NFL and NCAA football, and NCAA basketball for the STATS LLC reporter network. In addition to the more traditional baseball play-by-play information, the live baseball reports included more granular data such as broken bats, catcher blocks, first baseman scoops, and over a dozen distinct codes for balls and strikes.

Mr. Poserina took second place at the 2016 HIQORA High IQ World Championships in San Diego, California, finishing ahead of over 2,000 participants from more than 60 countries. He is a member of American Mensa, where he has served as a judge at the annual Mind Games competition that awards the coveted Mensa Select seal to the best new tabletop games.

Mr. Poserina has a B.A. in history and political science from Rutgers University. While studying there he called Scarlet Knight football, basketball, and baseball games for campus radio station WRLC.

J. Poserina

CHAPTER 1

How They Play the Game

There's an old saying: "It's not whether you win or lose; it's how you play the game." It comes to us from the 1908 poem "Alumnus Football" by the legendary sportswriter Grantland Rice, in which Coach Experience exhorts a young running back to keep on giving it his all against the defensive line of Failure, Competition, Envy, and Greed:

> Keep coming back, and though the world may romp across your spine,
> Let every game's end find you still upon the battling line;
> For when the One Great Scorer comes to mark against your name,
> He writes not that you won or lost but how you played the Game.

If we take "the Game" literally rather than as a metaphor, it's pretty easy to tell whether you have won or lost; there's usually a large electronic board standing by with that information at the ready. But quantifying "how you played the Game" is quite another matter, and it's something that has been evolving for centuries.

Among the earliest organized sports was cricket. Scorekeepers have been keeping track of cricket matches as far back as 1697, when two such reporters would sit together and make a notch in a stick for every run that scored. Soon the newspapers took an interest in the matches, and the recordkeeping gradually began to evolve. The earliest contest from which a scorecard survives where runs are attributed to individual batsmen took place between Kent and All England at London's Artillery Ground on June 18, 1744. In the early 19th century the accounts would include the names of bowlers when a batsman was dismissed. Fred Lillywhite, an English sports outfitter who published many early books on cricket, traveled with a portable press with which he both wrote newspaper dispatches and printed scorecards whenever he needed them. He accompanied the English cricket team, captained by George Parr, when it made its first barnstorming tour of the United States and Canada in 1859, served as scorer, and later published an account of the trip. Cricket had been played in North America since at least 1709; Benjamin Franklin brought home a copy of the 1744 rule book from his second trip to England, and over the first half of the 19th century teams were organized

Optimal Sports Math, Statistics, and Fantasy.
DOI: http://dx.doi.org/10.1016/B978-0-12-805163-4.00001-3
Copyright © 2017 Robert Kissell and James Poserina.
Published by Elsevier Ltd. All rights reserved.

in cities and colleges from Virginia to Ontario. The matches that George Parr's XI, as the English squad was known, played against local teams were modestly successful in raising interest for cricket in America, but those gains soon faded with the outbreak of the US Civil War. Soldiers on both sides tended to prefer cricket's young cousin, which was already establishing itself from the cities to the coal towns of the nation. By the time the English team returned for a second visit in 1868, cricket's popularity in America had declined. However, the connection between sports and statistics that had begun with cricket back in England was about to take on a completely new dimension.

Of all the organized American sports, it's perhaps not terribly surprising that the one game to make numbers a fundamental part of its soul would be baseball. Every event that occurs in a baseball game does so with few unknown quantities: there are always nine fielders in the same general area, one batter, no more than three base runners, no more than two outs. There is no continuous flurry of activity as in basketball, soccer, or hockey; the events do not rely heavily upon player formations as in football; there are no turnovers and no clock—every play is a discrete event. After Alexander Cartwright and his Knickerbocker Base Ball Club set down the first written rules for this new game on September 23, 1845, it took less than a month for the first box score to grace the pages of a newspaper. The *New York Morning News* of October 22, 1845, printed the lineups of the "New York Club" and the "Brooklyn Club" for a three-inning game won by Cartwright's team 24−4, along with the number of runs scored and hands out (batted balls caught by a fielder on the fly or "on the bound," the first bounce) made by each batter. Thus was born a symbiotic relationship: newspapers print baseball results to attract readers from the nascent sport's growing fan base, giving free publicity to the teams at the same time.

While he was not the first to describe a sporting match with data, "in the long romance of baseball and numbers," as Major League Baseball's official historian John Thorn put it, "no figure was more important than that of Henry Chadwick."[1] Chadwick had immigrated to the United States from England as a boy in the 1830s. Once across the pond, he continued to indulge his interest in cricket, at first playing the game and later writing about it for various newspapers in and around Brooklyn.

---

[1] Thorn, J., Palmer, P., & Wayman, J.M. (1994). The history of major league baseball statistics. In *Total Baseball* (4th ed.) (p. 642). New York, NY: Viking.

He eventually found himself working for the *New York Clipper*, a weekly newspaper published on Saturdays by Frank Queen. "The Oldest American Sporting and Theatrical Journal," as the *Clipper* billed itself, was "devoted to sports and pastimes—the drama—physical and mental recreations, etc." It did cover a wide variety of sports and games, from baseball ("Ball Play"), cricket, and boxing ("Sparring" or "The Ring"), to checkers, billiards, and pigeon shooting. It would also cover the newest play opening at the local theater, and the front page would often feature poetry and the latest installment of a fictional story, for which the paper would offer cash prizes. Eventually the *Clipper* would drop "Sporting" from its motto and focus exclusively on theater, cinema, and state fairs.

By 1856, Chadwick had been covering cricket matches for a decade for the *Clipper*. Returning from one such match, he happened to pass the Elysian Fields in Hoboken, New Jersey, where many of New York's and Brooklyn's teams would come to play, open space being much easier to come by on the west bank of the Hudson. "The game was being sharply played on both sides, and I watched it with deeper interest than any previous ball game between clubs that I had seen," he would later recall. "It was not long, therefore, after I had become interested in baseball, before I began to invent a method of giving detailed reports of leading contests at baseball."[2] Those detailed reports were box scores, and while he did not invent them he did much to develop and expand them.

On August 9, 1856, describing a game in which the Union Club of Morrisania, New York, defeated the New York Baltics 23–17, the *Clipper* published what was the forerunner of the line score, a table of runs and hands out with the hitters on the vertical axis and innings on the horizontal. The first actual line score would appear on June 13, 1857, for the season opener between the Eagles and the Knickerbockers, with each team's inning-by-inning tallies listed separately on its respective side of the box score. The line score would not appear in the *Clipper* again until September 5, describing a six-inning "match between the light and heavy weights of the St. Nicholas Base Ball Clubs on the 25th"[3] of August, this time listed vertically. In the following week's edition, Chadwick generated the first modern line score, horizontally with both teams together. The following season saw the first pitch counts. In an

---

[2] Quoted in Schwarz, A. (2004). *The Numbers Game: Baseball's lifelong fascination with numbers* (p. 4). New York, NY: St. Martin's Griffin.

[3] Light vs. heavy. *The New York Clipper*, *V*(20), September 5, 1857, 159.

August 7, 1858, game between Resolute of Brooklyn and the Niagara Club of South Brooklyn, won by Resolute 30—17 when play was halted after eight innings because of darkness, a whopping 812 pitches were thrown. Resolute's R.S. Canfield threw 359 pitches, including 128 in one inning, with John Shields of Niagara tossing 453. At the plate, both scored four runs. The lineup section of this box score still included just hands out (now called "hands lost" or just "H. L.") and runs. It would be the late 1870s before the box score would settle into a generally consistent format. Alfred H. Wright, a former player who later wrote for the Philadelphia *Sunday Mercury* before becoming the *Clipper*'s baseball editor, was the first to publish league standings, described in 1889 as "the checker-board arrangement now universally used to show the progress of the championship grade."[4]

In the December 10, 1859, edition of the *Clipper*, Chadwick generated a three-section season recap of the Brooklyn Excelsiors. "Analysis of the Batting" included games played ("matches played in"), outs made ("hands lost"), runs, and home runs for the 17 players, as well as the per-game means, maxima, and minima for hands lost and runs. The mean was formatted in the cricket style of average and over, meaning dividend and remainder; with 31 hands lost in 11 games (2 9/11), Arthur Markham's average and over was listed as 2—9. "Analysis of the Fielding" consisted of total putouts "on the Fly," "on the Bound," and "on the Base," as well as per-game maxima and the number of games in which no putouts of each type were recorded, followed by total number of games, total number of putouts, and average and over for putouts per game. "Additional Statistics" contained paragraphs of what we today might consider trivia: highest (62) and lowest-scoring (3), longest (3:55) and shortest (1:50), and earliest (May 12) and latest (October 25) games; the number of innings in which no runs were scored (29 of 133) and in which the Excelsiors scored in double figures (6). Chadwick acknowledged that there were gaps in his analysis, but nonetheless prevailed upon his dear readers to fear not:

> As we were not present at all the matches, we are unable to give any information as to how the players were put out. Next season, however, we intend keeping the correct scores of every contest, and then we shall have data for a full analysis. As this is the first analysis of a Base Ball Club we have seen

[4] Palmer, H. C., et al. (1889). *Athletic sports in America, England and Australia* (pp. 577—578). New York, NY: Union House Publishing.

published, it is of course capable of improvement, and in our analysis of the other prominent clubs, which we intend giving as fast as prepared, we shall probably be able to give further particulars of interest to our readers.[5]

The only table in the "Additional Statistics" section was a list of who played the most games at each position; Ed Russell pitched in eight of the Excelsiors' 15 games in 1859, and, strikingly from a modern perspective, that was the only time pitching is mentioned in the entire report. But at second glance, perhaps it's not so surprising that pitching was an afterthought. In the early days of baseball, the pitcher was the least important member of the team. He was required to throw underhand, as one would pitch horseshoes, which was how the position got its name. It also took some years before a rule change allowed him to snap his wrist as he delivered the ball. There were essentially two strike zones: the batter (or "striker" as he would be called) could request either a high pitch, meaning between the shoulders and the waist, or a low pitch, between the waist and the knees. If a striker requested a high pitch, and the pitcher threw one at mid-thigh level, the pitch would be called a ball, even if it went right over the middle of the plate. It took nine balls before the striker would be awarded first base, and a foul ball was no play, so strikeouts were rare; even so, Chadwick considered a strikeout as a failure by the batter and not a mark in favor of the pitcher, which it officially would not become until 1889. With so few opportunities for strategy, it would be the pitcher's job, more or less, to get the strikers to put the ball in play, which they did with great regularity. Fielding was much more of an adventure—the first baseball glove was still a decade away, and wouldn't become customary for two more after that; a single ball would last the entire game, regardless of how dirty or tobacco juice-stained or deformed it might become—so double-digit error totals were as common as double-digit run totals. Until 1865 some box scores even had a dedicated column for how many times each batter reached base on an error right next to the column for hits. As today, the statistical analysis reflected those aspects of the game considered most important.

But in another sense, statistics are separate from the game. We can change the way we record the events of the game and the significance we impart to them without changing the game itself. As Shakespeare might have put it, that which we call a walk by any other name would still place a runner on first base. Such was the case in 1887, when baseball's

---

[5] Excelsior club. *The New York Clipper*, *VII*(34), December 10, 1859, 268.

statistical rules were modified to consider bases on balls as base hits in order to artificially inflate batting averages. "Baseball's increasing audience seemed fascinated with batting stars," wrote the great sportswriter Leonard Koppett. "Batting stars were usually identified by their high batting averages, as the dissemination of annual and daily statistics continued to grow."[6] The batting average had been around since 1868—the *Clipper* analyses presented each player's "Average in game of bases on hits," doing so in decimals rather than average-and-over format. In 1876, announcing the conditions for its batting title, the National League adopted hits divided by at-bats as the definition of batting average and, for better or for worse, it has remained the go-to number for evaluating a hitter's prowess at the plate ever since.

Scoring walks as hits had an effect: batting averages leaguewide skyrocketed, though they received an additional boost from some other whimsical changes to the actual rules of play—pitchers could only take one step when delivering rather than a running start and, perhaps most glaring of all, it became one, two, three, four strikes you're out at the old ball game. Some sportswriters began referring to walks as "phantoms." Chadwick's *Clipper* was not impressed with "the absurdities of the so-called official batting averages of the National League for 1887" and refused to go along. "The idea of giving a man a base-hit when he gets his base on balls is not at all relished here. Base-hits will be meaningless now, as no one can be said to have earned his base who has reached first on balls."[7] Its pages considered the Phillies' Charlie Buffinton as having thrown two consecutive one-hitters, first shutting out Indianapolis 5−0 on August 6 and then having the Cubs "at his mercy, only one safe hit being made off him, and that was a scratch home-run"[8] in a 17−4 romp 3 days later; officially, with walks as hits though, they went down as a three-hitter and a five-hitter. Chadwick's bulletins gave two batting averages for each player, one "actual" and one "under National League rules." The differences were substantial: Cap Anson won the batting title over Sam Thompson, .421 to .407, but without the rule, Thompson would best Anson, .372 to .347. In 1968, baseball's Special Records Commission formally removed walks from all 1887 batting averages,

---

[6] Koppett, L. (1998). *Koppett's concise history of major league baseball* (p. 54). Philadelphia, PA: Temple University Press.

[7] From the hub. *The New York Clipper, XXXIV*(38), December 4, 1886, 603.

[8] Baseball: Chicago vs. Philadelphia. *The New York Clipper, XXXV*(No. 23), August 20, 1887, 361.

though Anson is still officially the 1887 National League batting champion. Overall, the National League hit a combined .321. Of players who appeared in at least 100 games, more hit over .380 than under .310. Throw out the walks, and it's a different story: the league average dips 52 points to .269, and the median among those with 100 games played plunges from .343 to .285. The story was the same in the American Association, whose collective average was .330, inflated from .273. Its batting crown went to Tip O'Neill of the Louisville Colonels, who was over .500 for much of the year before a late slump dragged him down to a paltry .485 (at the time listed as .492). Six batters with 100 games played broke .400, twice as many as were under .300. And after all of that, attendance remained more or less flat. In the face of considerable opposition to walks-as-hits, the Joint Rules Committee of the National League and American Association abandoned the new rule following the 1887 season, reverting the strikeout to three strikes as well. Despite the outcry, it's important to note that counting walks in the batting average did not change the game itself, only the way it was measured.

There were a few rare situations in which players and managers would allow statistics to influence their approach to the game. Chadwick, who by 1880 had spawned an entire generation of stat-keepers, was concerned about this: "The present method of scoring the game and preparing scores for publication is faulty to the extreme, and it is calculated to drive players into playing for their records rather than for their side." In the late 1880s, with batting averages dominating the statistical conversation, some hitters were reluctant to lay down a bunt—sure, it might advance a runner, but at what cost? They "always claimed that they could not sacrifice to advantage," *The New York Times* reported. "In reality they did not care to, as it impaired their batting record."[9] Scorekeepers started recording sacrifices in 1889 but would still charge the hitter with an at-bat; this was changed in 1893 but it would be a year before anyone actually paid attention to it and three more before it was generally accepted. Another such scenario is of much more recent vintage with the evolution of the modern bullpen. The save had been an official statistic since 1969, but as the turn of the century approached and relief pitching became much more regimented—assigned into roles like the closer, the set-up guy, the long man, and the LOOGY (lefty one-out guy, one of baseball's truly great expressions)—it became customary and even expected that the closer,

[9] Quoted in Schwarz, p. 19.

ostensibly the best arm in the pen, would only be called in to pitch with a lead of at least one but no more than three runs in the ninth inning or later. Sure, the game might be on the line an inning or two earlier with the bases loaded and the opposition's best slugger up, but conventional wisdom dictates that you can't bring the closer in then because it's not a "save situation." When the visiting team is playing an extra-inning game, the closer is almost never brought in when the score is tied, even though in a tie game there is no margin for error—if a single run scores the game is over—whereas in the "save situation" he can allow one, possibly two, or maybe even three runs without losing the game.

Over the years, other statistics would begin to emerge from a variety of sources. A newspaper in Buffalo began to report runs batted in (RBI) as part of its box scores in 1879; a year later the *Chicago Tribune* would include that figure in White Stocking player stats, though it would not continue that practice into 1881. In his *Baseball Cyclopedia*, published in 1922, Ernest J. Lanigan wrote of the RBI that "Chadwick urged the adoption of this feature in the middle [18]80s, and by 1891 carried his point so that the National League scorers were instructed to report this data. They reported it grudgingly, and finally were told they wouldn't have to report it."[10] By 1907 the RBI was being revived by the *New York Press*, and thirteen years later it became an official statistic at the request of the Baseball Writers Association of America, though that didn't necessarily mean that others were enamored of it. For one, *The New York Times* didn't include RBIs in its box scores for another eight years, when it placed them in the paragraph beneath the line score alongside other statistics like extra base hits and sacrifice flies. And it wasn't until 1958 that the *Times* moved the RBI into its own column in the lineup, where we expect it to be today.

Occasionally a new statistic would arise from the contestants themselves. Philadelphia Phillies outfielder Sherry Magee had a tendency to intentionally hit a fly ball to the outfield to drive in a runner on third, and his manager Billy Murray didn't think that an at-bat should be charged against Magee's batting average in such situations. His lobbying efforts were successful, as on February 27, 1908, the major leagues created a new statistic—this "sacrifice fly" would find itself in and out of favor and on and off the list of official statistics before finally establishing itself in 1954. Perhaps with an eye toward his own situation, in 1912,

[10] Lanigan, E. *Baseball cyclopedia*, p. 89.

Detroit Tigers catcher Charles Schmidt requested that scorers keep track of base runners caught stealing; before then, the only dedicated catching statistic was the passed ball.

Some statistics underwent changes and evolutions. In 1878, Abner Dalrymple of the Milwaukee Grays hit .356 to win the National League batting title, but only because at that time tie games were not included in official statistics. Had they been, Dalrymple would have lost out to Providence's Paul Hines, .358 to .354. They later would be, but unfortunately for Hines, this would not happen until 1968, 33 years after his death, when he was recognized as having won not only the 1878 batting title but also, by virtue of his 4 home runs and of the 50 RBIs for which he was retroactively credited, baseball's first-ever Triple Crown. When the stolen base was introduced in 1886, it was defined as any base advanced by a runner without the benefit of a hit or an error; thus a runner who went from first to third on a single would be considered to have advanced from first to second on the base hit, and to have stolen third. In 1898 this definition was revised to our modern understanding of a stolen base, and the notion of defensive indifference—not crediting an uncontested stolen base late in a lopsided game—followed in 1920. Also in 1898, the National League clarified that a hit should not be awarded for what we now know as a fielder's choice, and errors should not automatically be charged on exceptionally difficult chances.

The first inkling of the notion of an earned run dates back to 1867, when Chadwick's box scores began reporting separately runs that scored on hits and runs that scored on errors; even so, to him it was a statistic about batting and fielding, not pitching, much like he considered strike-outs about batting, not pitching, and stolen bases about fielding, not base-running. For the 1888 season, the Joint Rules Committee established that a "base on the balls will be credited against the pitcher in the error column."[11] The *Clipper* opined that "[t]here must be some mistake in this matter, as it will be impossible to make a base on balls a factor in estimating earned runs if the rule, as stated above, charges an error to the pitcher for each base on balls."[11] (This is not to say that Chadwick was necessarily on the right side of history about everything. In the same article, he refers to pinch hitters—"the question of each club having one or more extra men in uniform who may be introduced into the game at any time"—as a "dangerous innovation.") Chadwick resisted the notion of the earned

---

[11] Amending the rules. *The New York Clipper*, XXXV(36), November 19, 1887, 576.

run as a pitching metric until his death in 1908. In August 1912, National League Secretary John Heydler began recording earned runs officially and created a new statistic, the earned run average (ERA), which he defined not as earned runs per game, but as earned runs per nine innings. The American League would follow suit for the start of the 1913 season. As relief pitching became more of a factor in the modern game, the statistics had to be updated to reflect that. A pitcher's ERA, of course, reflects only earned runs and not unearned runs, which are runs that the team at bat scores by virtue of some sort of failure by the team in the field, namely a passed ball or an error other than catcher's interference, that they would not have scored otherwise. Whether a given run that scores after an error is committed is earned or unearned depends on the game situation. The official scorer is charged with reconstructing the inning as if the error had not occurred, using his best judgment with regard to where the base runners would have ended up, and to record as unearned any run that scored after there would have been three outs. Thus if a batter reaches on an error with the bases empty and two outs, any run that scores for the rest of that inning would be unearned, since the error obviated what should have been the third out. This was good news for a relief pitcher, as he could enter the game in that situation and allow any number of runs—they all would be unearned and his ERA would only go down. In 1969 the earned run rule was changed so that a reliever could not take advantage of the fact that an error had occurred before he entered the game. Now any run charged to his record during the inning in which he came on to pitch would be an earned run (barring, of course, another two-out error). One shortcoming of the ERA with respect to relief pitchers is that runs are charged to whichever pitcher put the scoring runner on base, not who allowed the runner to score. A relief pitcher entering a game with a runner on first base who serves up a two-run home run to the first batter he faces would only be charged with one run. The run scored by the man on first would be charged to the previous pitcher, even though he allowed the runner to advance only one base compared to three by the reliever.

With the evolution of all of these statistical categories, it wasn't long before writers and analysts began to use them prospectively rather than simply retrospectively, launching a science that continues today. Hugh Fullerton was a Chicago-based sportswriter who was unusual for including direct quotes from players and managers in his articles. He also was a careful observer of the game, and would achieve lasting fame for his statistical analyses of the Chicago White Sox in the World Series. In 1906, the first cross-town World Series pitted the White Sox against the Cubs, the dominating champions of the National League. The 1906 Cubs won 116 games, a record not matched until 2001 (although the 116–46 Seattle Mariners required seven more games to do it than the 116–36–3 Cubs), including an unparalleled 60–15 record on the road. Aside from the celebrated infield of Tinker, Evers, and Chance, the "Spuds," as they were sometimes called, were led on the mound by future Hall of Famer Mordecai "Three Finger" Brown, who went 26–6 with nine shutouts. They won the National League by 20 games over the New York Giants, and scored half a run more and allowed half a run fewer per game than anyone else. On the South Side, the White Sox and their respectable 93 wins edged the New York Yankees by just three games. Their offense scored 3.68 runs per game, just barely above the league average of 3.66, and their league-worst .230 team batting average and seven total home runs earned them the dubious nickname of "The Hitless Wonders." What chance could they have possibly had in the World Series against the most dominant ball club in history?

Plenty, thought Fullerton. "The White Sox will win it—taking four out of the first six games ... winning decisively, although out-hit during the entire series."[12] He acknowledged that the Cubs were "the best ball club in the world and perhaps the best that was ever organized ... [b]ut even the best ball club in the world is beatable." So sure of this was Fullerton that he promised to "find a weeping willow tree, upon which I will hang my score book and mourn" if his prediction didn't come true. His confidence stemmed from what he called "doping," a term that referred to something far different then than it does today. Fullerton described doping in a 1915 article published in the *Washington Times*: "We will first take every player on the two teams that are to fight for the championship, and study the statistics. We will find out what each man bats against right and left-handed pitchers, against speed pitching, and against slow pitching. We will figure his speed, aggressiveness, condition, and disposition." He would then take these analyses and meet with the players, their teammates, their opponents, and "men who know him better than I do."[13] He would then rate the players on batting, fielding, and value to his team on a scale of 0 to 1000, adding

---

[12] Fullerton, H.S. (October 15, 1906). Series verifies Fullerton's Dope. *The Chicago Tribune*, 4.

[13] Fullerton, H.S. (September 27, 1915). Hugh S. Fullerton explains his famous system of 'Doping' the Greatest Series of the Diamond. *The Washington Times*, 11.

or subtracting for factors such as defensive position or their physical and psychological state. "[Ty] Cobb's total value to his team in a season I figure at 850. Now, it makes no difference whether we say Cobb is 750 or 500 or 250. It is an arbitrary figure assumed for purposes of comparison."[13] Since there was no interleague play, Fullerton handicapped the batting matchups based on pitchers the hitters had faced who were similar to those they would be facing in the World Series. He also considered what we today would call "park factors."

Fullerton concluded that Brown was "the kind of pitcher the Sox will beat."[14] The Cubs' leading hitter at .327, third baseman Harry Steinfeldt, was "not that good a hitter. He never hit .300 in his life before, and probably never would again." He reckoned that the offensive disparity between the two teams was misleading because "[a]n American league batter is hitting against good pitching about six games out of seven," twice as often as a National Leaguer. When his analysis was complete, he predicted that the White Sox would win in six games. It wasn't actually printed until October 15, the day after the White Sox did indeed win the 1906 World Series in six games, because, the editor's note explained, a "man in authority refused to take a chance in printing Mr. Fullerton's forecast, but it has been verified so remarkably by the games as played that it is now printed just as it was written."[15] The White Sox did beat Three Finger Brown in two of his three starts. Steinfeldt hit .250 in the World Series, and never hit .270 for a season again, let alone .300. Fullerton would continue his World Series doping, and in 1919 would be instrumental in uncovering the Black Sox conspiracy.

Hugh Fullerton would be followed by others. Ferdinand Cole Lane had a scientific background, working as a biologist for Boston University and the Massachusetts Commission of Fisheries and Game before editing for many years a monthly called the *Baseball Magazine*. F.C. Lane, as he was known, was an early proponent of run expectancy and a fierce critic of batting average, calling it "worse than worthless"[16] as a statistic for evaluating prowess at the plate.[17] "Suppose you asked a close personal

[14] Fullerton. Series verifies Fullerton's Dope.

[15] Fullerton. Series verifies Fullerton's Dope (editor's note).

[16] Quoted in Schwarz, p. 34.

[17] In 1925, Lane published *Batting: One thousand expert opinions on every conceivable angle of batting science.* In an apparent about-face, he concluded the chapter "What the Records Tell Us" thusly: "To sum up, baseball owes a great deal to the records. Batting averages are the most accurate of these records. They serve as the only fair basis for comparing old time batters with modern hitters, or in comparing one present-day hitter with another." Lane, *Batting*, p. 14.

friend how much change he had in his pocket and he replied, 'Twelve coins,' would you think you had learned much about the precise state of his exchequer?" he wrote in the March 1916 issue. "Would a system that placed nickels, dimes, quarters and fifty cent pieces on the same basis be much of a system whereby to compute a man's financial resources? Anyone who offered such a system would deserve to be examined as to his mental condition."[18] Lane proposed not slugging percentage but a system of weights applied to singles, doubles, triples, home runs, and stolen bases in proportion to what his run expectancy analyses told him each event contributed to scoring a run. Singles, he calculated, were worth 0.4 runs, doubles 0.65, triples 0.9, home runs 1.15, and stolen bases and walks 0.25. Lane then applied this formula to batting statistics, and concluded that the Brooklyn Dodgers' two-time former National League batting champion Jake Daubert, whose .301 average was fifth best in 1915, was not as good a hitter as the Phillies' Gavvy Cravath, though Cravath only finished 12th at .285. Of Daubert's 151 hits, 120 (79.4%) were singles, compared to 87 of 149 (58.4%) for Cravath; he also trailed far behind Cravath in doubles (21−31) and especially in home runs (2−24). Lane's proposed batting average would divide those run values by at-bats; his model returned a total of 58 runs created by Daubert and 79 by Cravath, and averages of .106 and .151 respectively. "[T]he two are not in the same class," he concluded. "And yet, according to the present system, Daubert is the better batter of the two. It is grotesqueries such as this that bring the whole foundation of baseball statistics into disrepute."[19]

Branch Rickey, the brilliant baseball executive, also ventured into the realm of statistics. In a 1954 *Life* magazine article called "Goodby to Some Old Baseball Ideas," Rickey, then the general manager of the Pittsburgh Pirates, proposed what he called a "bizarre mathematical device" to measure a team's efficiency. "The formula, for I so designate it, is what mathematicians call a simple, additive equation."[20] The first part signified offense and the second pitching and defense, and the difference closely correlated a team's win−loss record: 1.255 for the first-place Brooklyn Dodgers and 0.967 for the last-place Pirates for the 1953 season. Despite its

---

[18] Lane, F.C. (March 1916). Why the system of batting averages should be changed. *The Baseball Magazine*, *16*(No. 5), 41−42.

[19] Lane, F.C. (March 1916). Why the system of batting averages should be changed. *The Baseball Magazine*, *16*(No. 5), 47.

[20] Rickey, B. (August 2, 1954). Goodby to some old baseball ideas. *Life Magazine*.

awkward appearance, the equation consisted of basic arithmetic and common statistics like hits, walks, at-bats, earned runs, and strikeouts:

$$G = \left(\frac{H + BB + HP}{AB + BB + HP} + \left(\frac{3(TB - H)}{4AB} \right) + \frac{R}{H + BB + HP} \right)$$
$$- \left(\frac{H}{AB} + \frac{BB + HB}{AB + BB + HB} + \frac{ER}{H + BB + HB} - \frac{SO}{8(AB + BB + HB)} - Fld\% \right)$$

Rickey claimed that his formula, developed in conjunction with Dodgers statistician Allan Roth and "mathematicians at a famous research institute," could reveal a team's strengths and weaknesses, why a team is showing improvement or regression, or which players are helping a team and which are not. He was persuaded that "[a]s a statistic RBIs were not only misleading but dishonest. They depended on managerial control, a hitter's position in the batting order, park dimensions, and the success of his teammates in getting on base ahead of him." He begrudgingly included fielding percentage so that defense would be represented, even as he called it "utterly worthless as a yardstick" and "not only misleading, but deceiving." Rickey was bringing mathematics into the front office half a century before Billy Beane and *Moneyball*. "Now that I believe in this formula, I intend to use it as sensibly as I can in building my Pittsburgh club into a pennant contender. The formula opened my eyes to the fact that the Pirates' [on-base percentage] is almost as high as the league-leading New York Giants. We get plenty of men on base. But they stay there! Our clutch figure is pathetically low, only .277 compared to New York's .397." The Pirates' clutch figure, which he defined as "simply the percentage of men who got on base who scored," would improve modestly to .301 by the end of 1954 and to .319 in 1955 when health problems forced Rickey to retire, leaving unresolved the question of whether his methodology would translate into long-term success.

Chadwick, Lanigan, Fullterton, Lane, Rickey, and others like George Lindsey and Earnshaw Cook represented small steps for the science of sports analytics, but it was a night watchman who made the giant leap. Not much typically happens in the wee small hours at a pork and beans factory outside of Kansas City, which afforded a man named Bill James plenty of time for examining the numbers of baseball. James's mind was bursting with ideas for different analyses of the 1976 season. But without a newspaper, and with only a few freelance articles in *Baseball Digest* to his credit, if he was going to put out a book he would need to do it himself. He spent hours and days and weeks in the quiet, dark factory crunching numbers by hand. He wrote the text out longhand and handed it over to his girlfriend for a rendezvous with a manual typewriter. Finally he placed a small classified ad in *The Sporting News* to inform the world of *The 1977 Baseball Abstract: Featuring 18 Categories of Statistical Information That You Just Can't Find Anywhere Else*, cover price $3.50. It sold 70 copies.

James self-published the *Abstract* again for 1978. In November, he was contacted by one of his readers, himself a writer. Dan Okrent, who would go on to write *Nine Innings* and become one of the founders of what evolved into fantasy sports, saw something in James and wanted to do a feature on him. While that was in the works, Okrent helped James secure the assignment of *Esquire*'s 1979 baseball preview. After more than two years, Okrent's piece on James finally ran in the May 31, 1981, issue of *Sports Illustrated*. Ballantine Books soon called with a book deal, and in 1982 the *Abstract* graduated from the ranks of self-publishing, and he continued to produce it until 1988.

In 1980 James coined the term *sabermetrics* to describe the science of baseball analytics in honor of the Society for American Baseball Research (SABR), which was founded in 1971. James has been pigeonholed as a "stat geek" by some members of the media and in the baseball establishment, both his critics and those who offer just a superficial description of him. James himself bristles at that characterization. "I would never have invented that word if I had realized how successful I was going to be," he later said. "I never intended to help characterize SABR as a bunch of numbers freaks."[21] James was much more than a mental spreadsheet sitting in the cheap seats; beyond the tables of figures, his *Abstracts* were full of commentary that could be as witty or as acerbic as it was insightful. Beyond baseball, James had a love of crime stories, and that helped inform his determination to get to the truth by way of evidence, to answer questions few have thought to ask. "There are millions of fans who 'know' that Clemente was better, because they saw him play, and there are millions who 'know' that Kaline was better for the same reason," he wrote in his book *Win Shares*. "I don't want to 'know' that way. I want reasons."[22]

Most of the analyses in the earlier *Abstracts* were based on end-of-season and career statistics or on information that could be gleaned from

---

[21] Gray, S. (2006). *The mind of Bill James* (p. 39). New York, NY: Doubleday.
[22] Gray, S. (2006). *The mind of Bill James* (p. 186). New York, NY: Doubleday.

box scores. A better understanding of baseball required lower-level data—actual play-by-play accounts. The problem was that in the 1980s, this information was simply not publicly available. The Elias Sports Bureau, founded 70 years earlier by Al Munro Elias and his brother Walter to provide timely and accurate statistics to New York newspapers, had become MLB's official statistician, and data beyond the box score was considered proprietary.[23] Beyond that, James argued that keeping this information secret was even an unfair labor practice, as it meant that during contract talks players had access to far less information than their counterparts across the negotiating table. So in 1984 James launched Project Scoresheet, which enlisted hundreds of volunteers to watch or listen to broadcasts and compile play-by-play data one game at a time using a special scorecard that James designed. Those accounts would be aggregated and the resulting dataset offered to the public, free of charge. In response, Elias published the *Elias Baseball Analyst*, containing data it had previously provided only to ball clubs along with a few arrows fired toward James, such as that their book contained "no arcane formulas with strange-sounding acronymic names" but rather represented "the state of the art for a variety of imitators, various individuals ... outside baseball [who] were fascinated by reports of its contents and attempted to copy it, with a notable lack of success."[24] For his part, James wrote that Elias had "ripped off my methods and my research so closely that many passages fall just short of plagiarism."[24]

James's goal was not simply to look at numbers for their own sake; he bristled at the volume of figures displayed on the screen during telecasts and the notion that the stats themselves were the sine qua non of the

---

[23] In 1995 Motorola began marketing a pager called SportsTrax that displayed live scores of NBA games. The NBA sued and received an injunction against Motorola, which was appealed. In 1997, the US Court of Appeals for the Second Circuit ruled that a while producing a live sporting event and licensing its broadcast were protected by copyright law, the "collection and retransmission of strictly factual material about the games" were not, as the scores and statistics arising from those games did not constitute "authored works." The Court also held that disseminating those facts did not constitute unfair competition: "With regard to the NBA's primary products—producing basketball games with live attendance and licensing copyrighted broadcasts of those games—there is no evidence that anyone regards SportsTrax or the AOL site as a substitute for attending NBA games or watching them on television." The raw data that the sports analytics community required was now public domain. The case was *National Basketball Association and NBA Properties, Inc. v. Motorola, Inc.*, 105 F.3d 841 (2d Cir. 1997).

[24] Quoted in Gray, p. 74.

game. Sure, Lenny Randle may have hit .285 with one RBI against lefthanders during night games on Thursdays in 1979, but did that necessarily mean anything? Baseball stats were not an end unto themselves but a means to an end, a tool to discover the fundamental theories underlying the game that had long lain hidden and to reveal those that were widely accepted but perhaps shouldn't have been. Baseball had long been run by baseball men, those who had grown up with the game and played the game and who were guided by their gut, their own experience, and a fealty to the way things had always been done. James represented a new way to approach the game based on mathematical investigation and the scientific method. Earlier analysts like Chadwick and Cook and Fullerton and Rickey and Roth had used numbers retrospectively, to describe players and teams based on things they had already done, and to predict how they were likely to perform against other players and teams. James was using evidence to argue that much of the conventional wisdom surrounding baseball was simply wrong. Teams would reach higher levels of success if they would only embrace at least his methods if not his conclusions. But James hadn't played baseball. He was an outsider preaching heresy, and by and large baseball did not take kindly to outsiders. They had no time for profound truths. By 1988, James was disillusioned with "the invasion of statistical gremlins crawling at random all over the telecast of damn near every baseball game"[25] along with the increasing amount of time he spent sending "Dear Jackass" letters to his most vitriolic critics. He walked away, at least for a while. "The idea has taken hold that the public is just endlessly fascinated by any statistic you can find, without regard to whether it means anything," he wrote. "I didn't create this mess, but I helped."[25]

But the sabermetric revolution was well underway. Bill James had lit a spark in countless imaginations, and thanks in part to him, raw data was now much more readily available. The third leg of the triangle was just as important: the rise of the home desktop computer along with spreadsheet software and eventually the World Wide Web. One of those imaginations inspired by James belonged to a Harvard graduate who had read the Bill James *Abstracts*—unusual perhaps but hardly unique. But this Harvard graduate was Sandy Alderson, who was also the general manager of the Oakland Athletics. Through *The Sinister First Baseman*, a collection of

[25] Bill James. *1988 Baseball Abstract*. Quoted in Lewis, M. (2003). *Moneyball: The art of winning an unfair game* (p. 95). New York, NY: W.W. Norton & Company.

writings by a sometime—National Public Radio contributor named Eric Walker, Alderson made a discovery. "For more than a hundred years, walks were made by hits," wrote Alan Schwarz in *The Numbers Game.* "Walker turned that strategy backwards and claimed that runs were scored not by hits, but by avoiding outs."[26] Outs were finite; you get only 27 of them in a regulation game. The best way to avoid making an out is to get on base. Unlike batting average, which is concerned only with hits, on-base percentage credits the batter for getting on base via a hit as well as drawing a walk or getting hit by a pitch. F.C. Lane had illustrated the concept of slugging percentage by rhetorically asking the total value of the coins in a friend's pocket rather than simply how many there were. Appreciating the importance of walks was like asking the friend to include the coins he had in his other pocket. Alderson hired Walker to consult for the Oakland front office. In 1982 Walker provided a paper explaining how the Athletics could improve their on-base percentage. In future years his reports would offer advice on off-season roster changes.

At the end of spring training in 1990, after he failed to make the big club, a former first-round draft pick who had stepped to the plate only 315 times over parts of six big-league seasons walked into Alderson's office. Just shy of his 28th birthday, he stated that he had had enough, and that rather than spend another season shuttling back and forth between the benches of Triple-A and The Show, he had an unconventional idea for how he could help the ball club. Billy Beane wanted to become an advance scout. By 1993 Beane had become assistant general manager. After the Athletics' owner died in 1995, the new ownership ordered the team to cut payroll. Alderson was going to use sabermetrics. He introduced Beane to the works of Bill James and Eric Walker, and Beane became a convert to objective rather than subjective analysis. Alderson left Oakland in 1997 for a job in the commissioner's office, and Beane succeeded him as general manager. He hired Paul DePodesta, a 25-year-old Cleveland Indians staffer, as his assistant GM and protégé. The two of them carried the banner of sabermetrics not only to use as a tool for evaluating players but to change the culture of the organization. Beane encountered substantial resistance from his scouting department, comprised largely of baseball lifers who could tell a ballplayer when he saw him and had no time for calculators and spreadsheets. Eventually, through a combination of epiphany and turnover, they came over to his side.

[26] Schwarz, p. 217.

The secret to Beane's success was what Alderson had taught him: to recognize which statistics were most closely correlated with scoring runs and winning games, and at the same time were undervalued if not overlooked by the rest of the baseball establishment. If Player A made $3 million and hit .290, and Player B made $650,000 but hit only .250, but both had similar on-base percentages, most other teams would ignore B and sign A. Beane would happily sign B and get nearly the same production at a fraction of the price. And if B was guilty of the crime of not "looking like a ballplayer," so much the better. And should he lose a productive player, as he did when Jason Giambi became a free agent and signed a contract with the New York Yankees, Beane realized that he could still get the same production out of his lineup by restoring Giambi's numbers, and that doing so did not necessarily require him to find an individual ballplayer to do so, if he could even find one who was available and affordable at all. With three new players whose total production was above average, Beane could replace the lost superstar as well as two underperforming marginal players at the same time while also cutting his already meager payroll. Once he had the players, he and his farm system had to get them to change their approach. A surefire way to avoid outs was to stop giving them away through sacrifices and ill-advised stolen base attempts. To maximize on-base percentage, hitters must look to draw more walks. (Ironically, Beane walked only 11 times in his playing career, putting up a lifetime on-base percentage of just .246 alongside a .219 batting average.) With one of the lowest payrolls in the major leagues the Athletics reached the playoffs five times between 2000 and 2006, topping the 100-win plateau twice.

A large component of Moneyball, as the strategy employed by Oakland's front office would be known, was that Beane was the only one doing it. He could get good players for cheap only so long as other teams undervalued them. But other ball clubs gradually learned from Oakland's example. Soon there would many more suitors vying for Player B's services, and his price wouldn't stay a bargain for long. In 2003 Bill James finally got his foot in the baseball establishment's door when he joined the Boston Red Sox as a senior consultant. After losing Game 7 of the American League Championship Series to the New York Yankees, the Red Sox fired manager Grady Little and replaced him with Terry Francona. Francona was receptive to the concept of analytics; he would at least consider them when making decisions in the dugout as he guided the Red Sox to their first World Series championship in 86 years.

Analytics departments began to appear in front offices around the major leagues. In 2016 they would reach a regrettable and perhaps inevitable milestone, one better suited for other sections of the newspaper: Chris Correa, the St. Louis Cardinals' scouting director who had started with the team as a data analyst, pleaded guilty to federal charges of hacking into the baseball operations database of the Houston Astros and was sentenced to 46 months in prison.

Today Alderson is the general manager of the New York Mets. In October 2015 Beane was promoted from general manager to executive vice president of the Athletics. DePodesta would later serve as general manager of the Los Angeles Dodgers and vice president of the Mets and the San Diego Padres. In January 2016 he left baseball for the NFL, becoming the chief strategy officer of the Cleveland Browns.

Football's past is even more gradual and evolutionary. Games that place some number of people on a field of some size with the object of advancing a ball over a goal line have been around for nearly as long as there have been civilizations. The ancient Greeks called such a game *harpaston*, and of our modern games it most closely resembled rugby. Harpaston was later adopted by the Romans as *harpastum*, and from there it spread to Britain, although it is generally accepted that its migration across the Channel likely did not take place as part of the Roman invasion. Around 1175, William Fitzstephen described how the people of London would engage in the "well known game of play ball after dinner on Shrove Tuesday,"[27] a tradition that would continue until 1830. Soon though the game would receive attention from the monarch, and not in a good way; King Edward II prohibited football in 1314 after complaints that it disturbed the peace, and his son Edward III would likewise declare a ban in 1349 on the grounds that playing "various useless and unlawful games"[27] like football distracted from the practice of archery. Across the Tyne in Scotland, James II appreciated the beneficial effects of the games people play, but nonetheless in a 1457 edict, famously quoted in a dissenting opinion by US Supreme Court Justice Antonin Scalia,[28] banned both football and golf, the former for its violence and brutality and the latter, like Edward III, for its effect on

---

[27] Baker, L.H. (1945). *Football: Facts & figures* (p. 4). New York, NY: Farrar & Rinehart.
[28] *PGA Tour, Inc. v. Martin*, 532 U.S. 661, 700 (2001) (Scalia, J., dissenting)

archery training. Shakespeare referred to football in both *A Comedy of Errors*[29] and *King Lear*.[30]

The word *football* itself referred originally not to a game where a ball is kicked by a foot, but rather to a family of games played on foot as opposed to on horseback. By the 19th century football had become the sport of schools and universities. With no central authority at the time, each school played a different variation depending on its surroundings. At schools like Charterhouse, where there wasn't a lot of room, the ball could be advanced only by kicking, whereas at places like Rugby, which had large fields, other means were allowed. It was there in 1823 that a game took place involving a student named William Web Ellis. Ellis caught a high kicked ball, and the rules at that time allowed him to attempt a drop kick. He was too far away to score with a drop kick, and with the light fading and the 5:00 bell nearly at hand, Ellis determined his only chance was to run the ball. His opponents were stunned and enraged at his audacity, which they attempted to punish by tackling him. Ellis had unwittingly invented the game that would be known simply as rugby. The Charterhouse and Rugby games would eventually become distinct, and formal rules would be drawn up for the respective sports by the Football Association in 1863 and the Rugby Football Union in 1871. The term "Association football" would soon be shortened to "soccer."

Football came across the Atlantic with the English colonists in the early 17th century. As the first colleges were established in the colonies that would become the United States, football was uncommon; Yale historian L.H. Baker ascribed this to the predominantly theological nature of American colleges, which frowned upon spending time playing sports when it could otherwise be used in the study of Scripture. But this would change. The earliest account of a college football game was in the *Harvard Register* of October 1827; its popularity at Harvard grew during the following decade. By 1840 it had spread to Princeton and Yale, two schools that both would ban it for its violence within 20 years. The first intercollegiate game took place in New Brunswick, New Jersey, on November 6, 1869, between Rutgers and Princeton. Rutgers challenged their rivals to a football game to avenge a 40−2 defeat in baseball, and prevailed 6−4 in a game that was not exactly football, not exactly soccer. Princeton won the rematch 8−0 on its home field and under its own rules with which the

[29] Act II, scene 1.
[30] Act I, scene 4.

Rutgers squad was unfamiliar, but the rubber game was canceled. The following year Rutgers defeated Columbia and lost to Princeton again. Football took a year off in 1871 after the violence of the game left a bad taste in many mouths, but in 1872 it spread to Yale, New York University, City College of New York, and Stevens Institute of Technology; football was on its way. The American Intercollegiate Football Association was formed in 1876 between Yale (which would quickly leave), Harvard, Princeton, and Columbia, and the rules it adopted were much like rugby, although punting dates back to this meeting as well. By the turn of the 20th century, football's violent nature dominated its virtues in people's minds yet again, and it got to the point that President Theodore Roosevelt threatened to outlaw the game. This led directly to the legalization of the forward pass and the formation of what is now the National Collegiate Athletic Association (NCAA) in 1906.

Unlike baseball, football's relationship with numbers was a long time coming. For many years, records were kept independently with no guarantee as to their completeness or accuracy. Numbers in football were rudimentary: points scored, wins and losses, yardage on rushing and total offense. Longest runs had been kept since 1884; tops on that list was a 115-yard run by Terry of Yale in an 1884 game against Wesleyan back when the field was still 110 yards long. Punting was recorded in terms of average yardage per punt since its adoption in 1876. Kickers were evaluated on field goals since 1873 and points-after-touchdown as early as 1906. But still there was no official statistician above the school level. This system finally changed in 1937 when Homer S. Cooke founded the American Football Statistical Bureau, marking the beginning of the modern era for college football statistics. To put this in context, by 1937 Babe Ruth, legendary for his marks of 714 career home runs and 60 in one season, was already retired. The first official statistics for rushing were simply games played, number of carries, and total net yardage. It would not be until 1970 that college football officially kept rate statistics for offense: yards per game for rushing, receiving, and total offense; catches per game for receiving and intercepting. Postseason games did not count in official statistics until 2002.

In 1979, the NCAA introduced the passer efficiency rating as a means to evaluate quarterbacks beyond touchdown passes and total yards. Statistics were aggregated from a 14-season period (1965–78) during which passers averaged 6.29 yards per attempt, and of those attempts 47.14% were completions, 3.97% were for touchdowns, and 6.54% were intercepted. These figures were incorporated into a formula such that a quarterback with those average statistics would have a rating of 100, and a touchdown pass was worth two interceptions:

$$PR = \frac{8.4 \cdot YDS + 100 \cdot COMP + 330 \cdot TD - 200 \cdot INT}{ATT}$$

To qualify, a passer must average 15 attempts per game and play in 75% of his team's games. The average passer rating across all of the NCAA's Football Bowl Subdivision (FBS; formerly known as Division I−A) has steadily grown over the years, from 104.49 in 1979 to 127.54 in 2006. The National Football League (NFL) also has a passer rating but it differs from the NCAA's. Adopted in 1973, the NFL's also looks at the same statistics and the formula is indexed to a 10-year aggregate sample, but the weightings and thus the scale are quite different. Unlike the NCAA's original average rating of 100, the NFL's average was 66.7, a number that has also gradually tracked upwards and is now well over 80. The NFL also limits each of the four categories to a minimum of 0 and a maximum of 2.375, thus the pro passer rating has an upper limit of 158.3.

$$C = 0 \le 0.05 \left(\frac{100 \cdot CMP}{ATT} - 30 \right) \le 2.375$$

$$Y = 0 \le 0.25 \left(\frac{YDS}{ATT} - 3 \right) \le 2.375$$

$$T = 0 \le 20 \left(\frac{TD}{ATT} \right) \le 2.375$$

$$I = 0 \le 2.375 - 25 \cdot \frac{INT}{ATT} \le 2.375$$

$$PR = \frac{C + Y + T + I}{0.06}$$

Both the NFL and NCAA refer to these figures as passer ratings, not quarterback ratings, as they do not encompass all of the aspects of the position such as rushing, play-calling, or avoiding sacks.

With no official statistics in earlier years, there was no way to objectively rank players against each other. Baseball had its batting averages, but what did college football have? Thus college football turned from objective to subjective means: the poll and the All-America Team. The first All-America Team dates back to 1889 and was compiled by Walter Camp, perhaps the most prominent of modern football's founding fathers, and possibly in collaboration with Caspar Whitney of *Week's Sport*. Soon it seemed that everyone wanted to jump aboard the All-America Team bandwagon; by 1909 the *Official Football Guide* contained 35 of them. "The popularity of the All-America Team spread quickly as the sport itself grew more popular and it became part of the duty of a sports writer to get up one of his own for publication," wrote L.H. Baker. "Ex-coaches were invited to make up lists for newspapers and other publications."[31] But the All-America Team was just a list of players from the late season. Who would make up a team of the best players ever? Enter the All-Time Team. The first such list appeared in the *New York Evening World* in 1904, and it tells us something about player evaluation at that time. The "All-Time All-Player" list was comprised entirely of gridders from Harvard, Yale, and Princeton, along with one from Army, each a northeastern school. Camp's All-Time Team from 1910 reflected the fact that he had traveled more extensively and seen more games himself, as it included players from Michigan, Pennsylvania, and Chicago. These were soon followed by All-Eastern teams, All-Western teams, All-Conference teams, and All-State teams; individual schools even named All-Opponent teams. Football grew so popular that one couldn't narrow down the best to a squad of 11, which begat the second- and third-team All Americas. With the rise of the one-way player in the late 1940s, begun by Fritz Crisler of Michigan in a game against Army in 1945, there would of course be separate All-America teams for offense and defense, debuting in 1950.

Subjective means have also been a huge factor in announcing the season's national champion. Major college football is unique in that it is the only sport in which the NCAA itself does not recognize a national champion. Founded in 1906 primarily as a rules-making body, the NCAA awarded its first national title in 1921 to Illinois at the National Track and Field Championship. By the time of that first NCAA national championship, intercollegiate football was already over 50 years old. The stronger football schools had since organized themselves into conferences, the oldest of which were the Southern Intercollegiate Athletic Association, founded in 1894, and the Big Ten, formed as the Western Conference in 1896. Other conferences included the Missouri Valley Intercollegiate

---

[31] Baker, L.H. p. 142.

Athletic Association (1907, today the Missouri Valley Conference), the Pacific Coast Conference (1915, today the Pac-12), the Southwestern Athletic Conference (1920), and the Southern Conference (1921).

For most of the 20th century, major college football's postseason consisted of schools receiving invitations, or "bids," to participate in games called bowls, which are independently operated and thus not under the direct purview of the NCAA. The oldest of the college bowl games is the Rose Bowl, first held in 1902, in conjunction with the Tournament of Roses Parade. The parade was not in addition to the game; rather the game was held to attract more people to the existing parade. The Pasadena Tournament of Roses Association was founded in 1890 to organize a parade on New Year's Day to celebrate the beauty both of flowers and of weather in southern California at a time of year when many back east were huddled under blankets, watching inch upon inch of snow pile up outside. And come those easterners did, and predominately from the Midwestern states. To attract even more visitors, and the money they brought along with them, the Association sponsored the Tournament East—West Football Game featuring the University of Michigan, which trounced West Coast representative Stanford 49—0. Perhaps because of the lopsided score at the expense of the approximate home team, football was replaced by other sporting events until 1916, when the Rose Bowl Game (as it would be known beginning in the 1920s) became a permanent fixture on the college football calendar for January 1, or January 2 if New Year's Day fell on a Sunday. Once postseason football resumed in Pasadena, the Rose Bowl Game would feature a champion from the West Coast and another team chosen by the Tournament of Roses Association. That arrangement lasted until 1946, when the Association signed an unprecedented contract with the Big Ten and Pacific Coast Conferences. The Rose Bowl was joined in the 1930s by the Sugar Bowl (1935), Orange Bowl (1935), Sun Bowl (1935), and Cotton Bowl Classic (1937). There were at least 10 other different postseason bowl or all-star games held irregularly between 1907 and 1939, many of which were held only once or twice; some invited specific teams rather than conference champions and for that reason are no longer officially recognized.

Following the bowl games, which traditionally take place on or within a few days of New Year's Day, various organizations would either vote, poll, or crunch numbers to recognize a team as the national champion for that season. All of these bowl games and polls and rankings arose independently and spontaneously, without the imprimatur of the NCAA, to fill a need that the public had for postseason play and the naming of a national champion as recognized by some accepted authority, and this all took place before the NCAA even created what would become its showcase postseason event. The first college basketball game was played in 1896, but it wasn't until 1939 that the NCAA held the men's basketball tournament. Likewise in college basketball, nature abhors a vacuum, and the void had already begun to be filled by two independent postseason tournaments: the National College Basketball (later the National Association of Intercollegiate Basketball, or NAIB) tournament in 1937, and the National Invitation Tournament (NIT) in 1938. At least with basketball, the NCAA was not too far behind the competing tournaments: it caught up with, overpowered, and eventually bought out the NIT; and the NAIB would become the National Association of Intercollegiate Athletics, comprised chiefly of small colleges. But with football, if the NCAA had decided to create a football playoff around the same time it did likewise for basketball, there would have been a lot of resistance; the bowl system and conference structures were just too long established. As an organization with voluntary membership, the NCAA had no power simply to impose a playoff. Moreover, as matters relating to football have been of paramount importance in how the NCAA has operated over the years, its governance would become dominated by the very football powerhouses that would have the most to lose financially by abandoning the bowl games in favor of a postseason tournament.

The NCAA recognizes over 30 different systems that have been used to name a national champion over the years, including some that have retroactively evaluated old records long after the fact. One such model was the Dickinson System. In 1926, University of Illinois economics professor Frank Dickinson created a ranking system to crown a national champion, which he did for the next 14 years. In 1935, Dickinson named Southern Methodist University (SMU) as the national champion, even though SMU had been defeated by Stanford 7−0 in the Rose Bowl. The Associated Press (AP) first polled its members on college football rankings in 1934, and it became a permanent institution in 1936. That season, the AP named Minnesota as the national champion, even though it had lost to Northwestern and both schools finished with identical 7−1 records. Notre Dame was the first wire-to-wire AP champion in 1943, and four years later the Irish would go back and forth with Michigan at the top of the poll throughout the season, both finishing 9−0. After the championship was announced, Michigan went to the Rose Bowl and trounced USC 49−0, whereas it was Notre Dame's policy until 1969 to decline all bowl bids. Michigan's dominant victory meant nothing, as the AP poll did not consider postseason games, and the national championship went to South Bend.

In 1950 the United Press (UP) created its own poll, but one of coaches rather than of sportswriters. The first time the AP and UP polls differed in their choice of national champions was in 1954, when the AP selected Ohio State and UP named UCLA. With the Buckeyes from the Big Ten and the Bruins from the Pac-8, this seemed to make the Rose Bowl a de facto national championship game. It did not happen, as the Pac-8 had a rule at the time that did not allow one of its members to appear in the Rose Bowl two years in a row. Twice a split championship happened because the AP and UP had different philosophies. In 1957 the AP crowned Auburn as the national champion; the UP disqualified Auburn from its poll as the Tigers had been placed on probation by the Southeastern Conference (SEC) for paying cash to two high school players, and instead named Ohio State. This happened again in 1974 when Oklahoma was on probation for recruiting violations and banned from appearing in a bowl game. The Sooners came away with the AP's Number 1 ranking, whereas the UP, now called United Press International (UPI), dropped Oklahoma from its ballots entirely and proclaimed USC instead. In 1959 both polls went to Minnesota, who then lost in the Rose Bowl. Soon both polls would include postseason games, the AP starting in 1965 and UPI in 1974.

As the major conferences were contractually tied to individual bowls, it was rare that the polls' differing No. 1 choices or their No. 1 and No. 2 could face each other in a postseason game; they would have to be either from the right conferences or not affiliated with a conference at all. Alabama of the SEC was able to play independent Notre Dame in the 1974 Sugar Bowl, and in 1986 the Fiesta Bowl hosted two independents as Penn State defeated the University of Miami. It was more usual that split champions would play separate bowl games, as was the case in 1990 when AP champion Colorado (11−1−1) did not face UPI champion Georgia Tech (10−0−1). The following year, there were again two champions, both of which were undefeated: Miami (AP) and Washington (UPI). There was no chance for a national title game because Washington, as Pac-10 champion, was contractually obligated to face Michigan in the Rose Bowl. A showdown between two undefeated teams could not happen.

The second consecutive split championship was quickly followed by a new alliance between some of the major bowls and the college football powerhouses called the Bowl Coalition. The champions of the Big East and Big 8 (today the Big 12) would meet in the Orange Bowl, of the Atlantic Coast Conference (ACC) and SEC in the Sugar Bowl, and the champion of the Southwest Conference (SWC) and Notre Dame in the Cotton Bowl. The Rose Bowl, with its exclusive conference and television contract, was not a part of the Bowl Coalition. The national championship would be contested in the Fiesta Bowl, but only if the No. 1 and No. 2 teams came from the Big East, ACC, or Notre Dame. If the No. 1 or No. 2 team came from the Big Ten or Pac-10, a national championship game would not be held. It soon became clear that nothing much could conceivably come from the Bowl Coalition, and in 1994 it disbanded.

The new flavor of the month was a 1994 proposal for a six-team playoff involving Notre Dame and the champions of the Big Ten, ACC, Pac-10, Big 8, and SEC. At the same time, the NCAA announced it would begin a study of its own playoff. Cedric Dempsey, the new executive director of the NCAA, cautioned that a "play-off system must prove it won't lose money.... The bowls are proven to be profitable."[32] The Rose Bowl claimed a disproportionate share of those profits, comprising fully a third of all bowl game television revenues, so naturally the Big Ten and Pac-10 had no interest in such a plan. A poll in February 1994 of the College Football Coaches Association revealed that 70% of its Division I membership opposed a playoff. By June the study was abandoned because of a lack of interest.

In 1998, after the Rose Bowl contract with the Big Ten and Pac-10 expired, the six power conferences along with Notre Dame created the Bowl Championship Series (BCS). The impetus was yet another split national champion in 1997, this time Michigan and Nebraska. The BCS was not a championship series per se; the existing bowls would remain in place and one would be designated as the championship game on a rotating basis. "If fans want a playoff," said Big Ten commissioner Jim Delaney, "they can get it from the NFL and the NBA."[33] In a June 22, 1998, article in *The Sporting News* called "Bowl Championship Series is Better than a Tournament," Tom Dienhart and Mike Huguenin[34] wrote that "a one-and-done playoff like the one the NCAA runs for basketball crowns the best team less often than any system college football has ever employed." Unlike the weekly and even preseason polls, the first BCS

---

[32] Talk of I-A playoff moves along slowly. *The NCAA News*, January 19, 1994, 27.

[33] Quoted in Smith, R.A. (2001). *Play-by-play: Radio, television, and big-time college sport.* Baltimore, MD: Johns Hopkins University Press.

[34] Dienhart, T., & Huguenin, M. (June 22, 1998). Bowl Championship Series is better than a tournament. *The Sporting News.*

rankings for the inaugural season were not released until mid-November. The BCS formula consisted of team's positions in the AP poll, their average ranking in three computer models (Jeff Sagarin, Anderson-Hester, and *The New York Times*), along with a strength of schedule component and a penalty for a loss, with the Number 1 ranking going to the school with the lowest score.

The BCS methodology was constantly evolving. After the first year the computer ranking component was refactored as five new models were added (Billingsley, Dunkel, Massey, Matthews, and Rothman) with each school's lowest rating among them being dropped. In 2001 Dunkel and *The New York Times* were replaced by Wolfe and Colley as the BCS sought to diminish the role played by margin of victory in the rankings. In addition, a quality win factor was introduced. Once all the numbers were run, a team's BCS score would be reduced by 0.1 points for a victory over the No. 15 school, 0.2 points for defeating No. 14, and so on up to 1.5 points for beating the No. 1 team. The next year the quality win factor was tweaked to include only wins over the top 10, not the top 15. The game of musical chairs among the computer models continued as Matthews and Rothman were out; Sagarin dropped his margin of victory component entirely as did *The New York Times* in its return to the BCS fold. Gone as well were the quality win, strength of schedule, and win−loss record components.

Amidst grumblings from commentators and some fans about the computer rankings, the BCS formula underwent its biggest transformation in 2004 as once again two teams were proclaimed as national champions, despite the fact that the primary reason for the BCS's existence was to prevent that from happening. In 2003, USC was ranked No. 1 in the polls but the computer models helped place Oklahoma and LSU in the BCS championship game. Now the BCS formula was simplified to include three components: the AP poll, the ESPN/*USA Today* coaches' poll, and the computers. There would be six models (Sagarin, Anderson and Hester, Billingsley, Colley, Massey, and Wolfe); each team's highest and lowest rankings among them would be discarded and the remaining four averaged. BCS officials considered weighting the polls 40% each and the computer models 20% before deciding, as BCS coordinator and Big 12 commissioner Kevin Weiberg put it, "that the equal weighting approach was the simplest."[35]

[35] Greenstein, T. (July 16, 2004). BCS revises ranking system. *Chicago Tribune*.

But the AP had finally had enough of the complaints and controversies and reformulations. On December 21, 2004, the AP's associate general counsel George Galt sent the BCS a cease-and-desist letter, saying that the AP had never formally agreed to "assist [the BCS] in preparing its rankings," and that the BCS's "forced association of the AP Poll with BCS has harmed AP's reputation" and violated the AP's copyright. "BCS's continued use of the AP Poll interferes with AP's ability to produce the AP Poll and undermines the integrity and validity of the AP Poll. BCS has damaged and continues to damage AP's reputation for honesty and integrity in its news accounts through the forced association of the AP Poll with the BCS rankings. ... Furthermore, to the extent that the public does not fully understand the relationship between BCS and AP, any animosity toward BCS may get transferred to AP."[36] One option to replace the AP Poll was that of the Football Writers Association of America (FWAA), but they were only amenable to having their poll used in an advisory capacity. "We don't want to be part of a formula," said FWAA executive director Steve Richardson.[37] Left with the coaches' poll, which was heavily criticized because its balloting was secret, and the computer models, the BCS briefly flirted with the idea of a selection committee before appointing the Harris Interactive College Football Poll to replace the AP.

The formula and the results thereof were not the only controversy. Beyond selecting the teams for the championship game, the BCS was also responsible for selecting the teams that would play in the oldest and most venerable bowl games by determining those schools that automatically qualified and those that were eligible for the pool of teams from which the bowls themselves would award at-large bids. The method by which the BCS did this left half of the FBS on the outside looking in by granting automatic qualifying bids for BCS bowls to some conferences but not to others. Unless playing in the championship game, the conference champions of the ACC and (until 2006) the Big East would play in the Orange Bowl, the Big Ten and Pac-12 in the Rose Bowl, the Big 12 in the Fiesta Bowl, and the SEC in the Sugar Bowl, with the remaining slots in those bowl games being awarded as at-large bids. So if your school was in one of the Power Six, life was straightforward: win your conference, go to a BCS bowl. But for those in the nonautomatic qualifying

---

[36] The Associated Press. (December 21, 2004). AP sends cease-and-desist letter to BCS. *USA Today.*

[37] Greenstein, T. (January 5, 2005). BCS selection committee still on table. *Chicago Tribune.*

conferences—Conference USA, the Mid-American Conference (MAC), the Midwest Conference (MWC), or Sun Belt—it seemed the only way to get to a BCS bowl game was to buy a ticket.

In 2003, the mid-majors began to revolt. Tulane considered dropping out of the FBS altogether because "of the inequities and restrictions inherent in the NCAA system and the Bowl Championship Series alliance," wrote its president Scott Cowen in *The New York Times*. "Tulane is a member of Conference USA, which is not a part of the BCS. This means that Tulane and other non-BCS football teams have virtually no— or only limited—access to the highest-paying postseason bowls governed by the BCS alliance."[38] Popular demand convinced Tulane to remain in the FBS, but the message still resonated. Fifty university presidents from the non-BCS conferences formed the Coalition for Athletics Reform, which brought the term "antitrust" back to the sports pages. When the BCS operating agreement came up for renewal, NCAA president Myles Brand helped to negotiate a new deal for 2006 that improved postseason revenue sharing, added a fifth BCS bowl, and offered non-BCS conferences a path to the BCS bowls, though that path was tougher to navigate than the one traveled by the BCS conferences.

While a power conference team could receive an automatic bid by winning its conference championship game, a mid-major would have to finish in the top 12 in the BCS standings, or ranked between 13 and 16 and above a champion from one of the power conferences. But what if two mid-majors met those requirements? Unless they finished No. 1 and No. 2, the higher ranked of the two would receive the automatic bid. The other would have to hope it received an at-large bid, which itself required nine wins and a finish in the top 14 of the BCS standings. As a single conference could receive no more than two BCS bids (unless two of its nonchampions finished 1 and 2), it was possible that the top 14 might not yield 10 BCS bowl bids, in which case the field would expand to the top 18, and then the top 22, and so on until all 10 bids were awarded. Things were likewise inequitable for independent schools as among them only Notre Dame, by finishing in the top eight, could qualify for an automatic bid.

Under the old system, only one non-BCS school, 11–0 Utah in 2004, made it to a BCS bowl, where the Utes romped over Pittsburgh

---

[38] Cowen, S. (June 15, 2003). How Division I-A is selling its athletes short. *The New York Times*.

35−7 in the Fiesta Bowl. Following the adoption of the 2006 agreement there was at least one "BCS Buster," as mid-majors that qualified for BCS bowl games would be known, every year except 2011. Only one at-large bid ever went to a mid-major, the 13−0 Boise State squad of 2009 that defeated TCU in the Fiesta Bowl 17−10; all others were automatic qualifiers by finishing higher than a power conference champion. Boise State went undefeated as well in 2008 but had to settle for the Poinsettia Bowl, also against TCU.

Still it seemed that no matter what the BCS tried to do to improve things, popular resentment against it continued to grow, and the march to the inevitable finally reached its destination. On June 26, 2012, the BCS Presidential Oversight Committee, consisting of 12 university presidents and chancellors, sat in Washington's Dupont Circle Hotel and listened to a presentation in which the conference commissioners and Notre Dame's athletic director proposed a playoff. Just three hours later, it was approved. "I think some of it was just battle fatigue because the general sporting public never really embraced the current system, even though that system did a lot of good things for college football," ACC commissioner John Swofford said at the time. "I think the longer that went on, the more people realized we had to do something that was different and better."[39] The new system was called a "plus-one" model, meaning that only one game would be added to the schedule. Two of the six major bowls would be designated as semifinals, and the winners would meet the following week in the championship game. Gone were the polls and computer rankings; now a selection committee would meet and through a series of votes determine which four teams would advance to the playoff based on win−loss record and strength of schedule, with extra consideration for those that won head-to-head matchups and their conference championships. While the playoff may expand in the future, the current agreement will run through 2025.

Among the three major sports, basketball was the latest to the party. Professional baseball and intercollegiate football were already well established when Dr. James Naismith developed the 13 original rules for "basket ball" in Springfield, Massachusetts, during the winter of 1891−92. Like football, basketball was long a college sport before turning pro. In the early 1930s, a sportswriter named Ned Irish, convinced that the game was

---

[39] Schlabach, M. (June 27, 2012). Playoff approved, questions remain. *ESPN.com*.

outgrowing on-campus gyms, conceived the idea of scheduling college basketball games in large arenas. In December 1934 he arranged a very successful doubleheader at New York's Madison Square Garden with NYU hosting Notre Dame in the opener and St. John's taking on Westminster (Pennsylvania) College in the nightcap. The following season he put on eight more doubleheaders. Perhaps inspired by the growing popularity of the new postseason games, such as the Orange, Sugar, and Cotton Bowls, he helped to expand the doubleheaders into a new postseason playoff called the National Invitation Tournament (NIT). The first NIT, held in 1938, consisted of six teams including the champion Temple Owls.

Not to be outdone, the NCAA introduced a tournament of its own in 1939. Unlike the NIT, which invited whomever it wanted, the NCAA divided schools up into eight regions, and a selection committee, comprised primarily of coaches, extended bids to one school from each region. The committee was not bound to select conference champions; it could offer an invitation to a team it considered to be stronger despite having a worse record.

Of the two tournaments, the NIT for many years was considered the more prestigious. In the days before television, there was more media exposure afforded by playing in New York rather than, say, Evanston, Illinois. Not wanting to be left behind, the NCAA held its 1943 tournament in New York and returned every year but one through 1950, when City College of New York (CCNY) won both the NIT and the NCAA tournaments. The following year, however, CCNY was implicated in a point-shaving scandal that involved six other schools, including three in the New York area, and may also have had connections to organized crime. The NCAA never held the tournament championship in New York again, and didn't even schedule preliminary rounds in Manhattan until 2014. The NIT, though, was inextricably linked with the Big Apple; the connection that had once provided a sense of legitimacy now contributed to the tournament's decline, as the public associated college basketball in New York with corruption. As the NCAA's tournament overtook the NIT in popularity, it prohibited schools from accepting bids to play in more than one tournament. This ban also extended to schools hosting the NCAA tournament. Such was the case in 1970, when ACC regular season champion South Carolina lost in the finals of the conference tournament after star player John Roche suffered an ankle injury. After failing to make the NCAA tournament, the Gamecocks were forced to decline an NIT invitation because they had been designated as the host school for the NCAA East Regionals in Columbia, South Carolina. The NIT had become the de facto consolation tournament for those schools that failed to get an invitation to the Big Dance. In 2005 the NCAA finally acquired the NIT.

The Big Dance would continue to get bigger. The field for the 1951 tournament expanded to 16. For the first time, automatic bids were given to 11 conferences, and the remaining 5 at-large spots were earmarked for the eastern (3) and western (2) parts of the country. Some conferences, including the Ohio Valley and Mid-American, did not receive automatic bids. Two years later the bracket grew again, this time to 22, which necessitated first-round byes for some teams, specifically those from conferences with better records in previous years' tournaments. The byes went away in 1975 as the tournament expanded again, this time to 32, but would return 4 years later when eight more slots were added, with another eight coming in 1981.

Seeding the tournament was a complicated process, as the bracket was divided into four regions and a 1979 rule limited each 10-team region to no more than two teams from any one conference. So in 1981 the NCAA devised a formula called the Rating Percentage Index (RPI) "to provide supplemental data for the Division I Men's Basketball Committee in its evaluation of teams for at-large selection and seeding of the championship bracket."[40] That it was intended to be supplemental is a point that the NCAA was sure to emphasize. "The committee is not unequivocally committed to selecting the top 48 teams as listed by the computer," said committee chairman Wayne Duke.[41] A February 1995 edition of the NCAA's official newsletter wrote that "[s]ince the RPI's invention the committee has come to rely on it less than when it was first created."[42]

The formula has been changed a few times since it was introduced, but originally it was comprised of four factors:

- Team success (40%): winning percentage in Division I games minus 0.01 for each non–Division I game the school played
- Opponents' success (20%): an unweighted average of the winning percentages for each of the school's opponents in Division I games, excluding games between the school and the opponent

---

[40] NCAA Division I men's basketball championship principles and procedures for establishing the bracket. *NCAA Sports.com*, April 11, 2005.

[41] Damer, R. (January 23, 1981). In all fairness. *The Chicago Tribune*.

[42] Bollig, L.E. (February 8, 1995). FYI on the RPI. *The NCAA News*, 1.

- Opponents' strength of schedule (20%): an unweighted average of each opponent's opponents' success factor
- Road success (20%): an unweighted average of the school's winning percentage in Division I road games and the opponents' success factor for those road opponents

The results of the RPI were kept confidential, and were not provided to schools until 1992. The NCAA did not release weekly RPI rankings until 2006. Before then, the RPI rankings in various media outlets were estimates of what the real RPI figures were. Rumors abounded that there was some sort of secret adjustment that made the media estimates inaccurate, and indeed there was. From 1994 until 2004, a bonus was added if more than half of a school's nonconference opponents were in the RPI's top 50, and a penalty assessed if more than half were outside of the top 150. There were additional bonuses for victories over top 50 and penalties for losing to sub-150 or non-Division I opponents. These adjustments "could move some teams up or down as many as five spots," said NCAA statistics coordinator Gary K. Johnson.[42] In 2005 these adjustments were replaced by coefficients applied to home and road games. Wins at home and losses on the road were multiplied by 0.6, road wins and home losses by 1.4. Neutral site games still counted as 1.0, win or lose.

Today the RPI formula is simpler. Winning percentage counts 25%, opponents' winning percentage 50%, and opponents' opponents' winning percentage 25%. Simpler, though, does not necessarily mean better. Like previous incarnations (as well as BCS computer models), the RPI reflects the NCAA's antipathy toward considering margin of victory, because of both the legacy of point-shaving scandals (CCNY in the 1950s and later at Boston College in the 1970s) and of a desire not to incentivize teams to run up the score. As secrecy is no longer an issue, the primary criticism of the RPI is that it is not an evaluation of a team so much as its schedule, stacking the deck in favor of schools from power conferences at the expense of those from mid-majors. "The RPI, I was certain, explained everything," wrote Dick Jerardi in The Philadelphia Daily News. "I really believed that. I was wrong. The RPI, it turns out, is little more than a brainwashing tool used on those of us who want easy answers. I wanted easy answers. After watching how these RPI numbers are used, it has become clear they are just an easy way out, something to justify the status quo."[43]

---

[43] Jerardi, D. (January 23, 2002). It's high time for the NCAA to KO the RPI. The Philadelphia Daily News.

Fully three quarters of the RPI figure is based on its opponents' records. In 2002 Pittsburgh defeated Rutgers in a January game by a score of 66—58; because the Panthers walked out of Piscataway with a record of 15—1, the Scarlet Knights actually moved up three spots in the RPI rankings.[44] A school in a mid-major conference having a breakout year is hampered by the lower winning percentages of its conference opponents. In 2001, Kent State went 23—9, winning both the Mid-American Conference's East division title and its tournament, garnering the MAC's automatic bid to the NCAA tournament as a No. 13 seed. The rest of the MAC, however, combined for a record of 160—190, including Western Michigan (7—21), Northern Illinois (5—23), Buffalo (4—24), and Eastern Michigan (3—25). Owing to this light conference schedule, the Golden Flashes' RPI ranking was 93. Rutgers, which went only 11—16 in the much tougher Big East (the rest of which was a combined 241—150), had an RPI ranking of 80. To increase its RPI, a mid-major school would be well served by scheduling tougher nonconference opponents even at the expense of its own win—loss record. A CNN/*Sports Illustrated* story from 2000 discussed so-called RPI consultants, advising institutions on how to game the system. As beefing up its nonconference schedule is not always an option, a school would almost have to win its conference's automatic bid if it had any hopes of earning a ticket to the dance. Of the 173 at-large selections from 2007 through 2011, only 31 (17.9%) went to schools outside of the six power conferences (ACC, Big East, Big Ten, Big 12, SEC, and Pac-10).

Though it is said to be just a tool, the RPI does play a concrete role in the tournament selection process, which is guided by a document called "NCAA Division I Men's Basketball Championship Principles and Procedures for Establishing the Bracket." The committee is charged with selecting the best at-large teams from all conferences. While committee members are permitted to vote for any schools they deem worthy up to 34, the ballots and information packs provided to them only encompass teams in the RPI's top 105. As the committee is narrowing down the number of candidates, the chair can request what the NCAA calls the "nitty-gritty" report, which among other things lists a team's overall, conference, and nonconference RPI rankings and its record against opponents ranked 1—50, 51—100, 101—200, and 201 or below.

Today, in the information age, the relationship between sports and numbers is closer than ever before. No longer must analyses be run by

---

[44] Wetzel, D. (January 9, 2002). RPI continues to lack an ounce of reason. *CBS Sportsline*.

collecting statistics from a newspaper, aggregating them with a pocket calculator, and reporting them with a manual typewriter as Bill James did in the late 1970s. Raw data is easily accessible, and anyone with Internet access, a spreadsheet application, and a little imagination can reveal those aspects of sports that had lain hidden for so long.

BIBLIOGRAPHY

Altham, H. S. (1926). *A History of Cricket*. London: George Allen & Unwin Ltd.

Associated Press. (December 21, 2004). AP sends cease-and-desist letter to BCS. *USA Today*.

Baker, L. H. (1945). *Football: Facts & figures*. New York, NY: Farrar & Rinehart.

bcsfootball.org. (n.d.).

bigbluehistory.net. (n.d.).

Bollig, L. E. (February 8, 1995). FYI on the RPI. *NCAA News*, 1.

Camp, W. (1896). *Football*. Boston, MA: Houghton & Mifflin.

Chadwick, H. (September 5, 1857). Light vs. heavy. *The New York Clipper*, 159.

Chadwick, H. (December 10, 1859). Excelsior club. *The New York Clipper*, 268.

Chadwick, H. (December 4, 1886). From the hub. *The New York Clipper*, 603.

Chadwick, H. (August 20, 1887). Baseball: Chicago vs. Philadelphia. *The New York Clipper*, 361.

Chadwick, H. (November 19, 1887). Amending the rules. *The New York Clipper*, 576.

collegerpi.com. (n.d.).

Cowen, S. (June 15, 2003). How Division I-A is selling its athletes short. *The New York Times*.

Damer, R. (January 23, 1981). In all fairness. *The Chicago Tribune*, Section 4, 9.

Dienhart, T., & Huguenin, M. (June 22, 1998). Bowl Championship Series is better than a tournament. *The Sporting News*.

Dinich, H. (June 27, 2012). Playoff plan to run through 2025. *ESPN.com*.

Douchant, M. (March 27, 1982). Basketball bulletin. *The Sporting News*, 16.

Fullerton, H. S. (October 15, 1906). Series verifies Fullerton's Dope. *The Chicago Tribune*, 4.

Fullerton, H. S. (September 27, 1915). Hugh S. Fullerton explains his famous system of "Doping" the greatest series of the Diamond. *The Washington Times*, 11.

Gray, S. (2006). *The Mind of Bill James*. New York, NY: Doubleday.

Greenstein, T. (July 16, 2004). BCS revises ranking system. *The Chicago Tribune*.

Greenstein, T. (January 5, 2005). BCS selection committee still on table. *The Chicago Tribune*.

Jerardi, D. (January 23, 2002). It's high time for the NCAA to KO the RPI. *The Philadelphia Daily News*.

Koppett, L. (1998). *Koppett's Concise History of Major League Baseball*. Philadelphia, PA: Temple University Press.

Lane, F. C. (March 1915). Why the system of batting averages should be changed. *The Baseball Magazine*, 41−47.

Lane, F. C. (2001). *Batting: One thousand expert opinions on every conceivable angle of batting science*. Cleveland, OH: Society for American Baseball Research.

Lanigan, E. J. (1922). *Baseball Cyclopedia*. New York, NY: Baseball Magazine Co.

Leckie, R. (1973). *The Story of Football*. New York, NY: Random House.

Lewis, M. (2003). *Moneyball: The art of winning an unfair game*. New York, NY: W.W. Norton & Company.

Maxcy, J. G. (2004). The 1997 restructuring of the NCAA: A transactions cost explanation. In J. Fizel, & D. F. Rodney (Eds.), *Economics of College Sports.* Westport, CT: Praeger Publishers.

National Collegiate Athletic Association. (February 15, 1981). Computer to aid in basketball championship selection. *NCAA News*, 4.

National Collegiate Athletic Association. (January 19, 1994). Talk of I-A playoff moves along slowly. *NCAA News*, 27.

National Collegiate Athletic Association. (April 11, 2005). NCAA Division I men's basketball championship principles and procedures for establishing the bracket. *NCAA Sports.com.*

National Collegiate Athletic Association. (2007). *2007 NCAA Division I Football Records Book.* National Collegiate Athletic Association.

Nest, D. S., et al. (1989). *The Sports Encyclopedia: Pro Basketball.* New York, NY: St. Martin's Press.

Office of the Commissioner of Baseball. (2015). *Official Baseball Rules: 2015 Edition.*

Palmer, H. C., et al. (1889). *Athletic Sports in America, England and Australia.* New York, NY: Union Publishing House.

Peeler, T. (March 15, 2000). Coaches miffed by Virginia Snub. *CNN/SI.com.*

Rickey, B. (August 2, 1954). Goodby to some old baseball ideas. *Life Magazine.*

rpiratings.com. (n.d.).

Schlabach, M. (June 27, 2012). Playoff approved, questions remain. *ESPN.*

Schwarz, A. (2004). *The Numbers Game: Baseball's lifelong fascination with numbers.* New York, NY: St. Martin's Griffin.

Smith, R. A. (2001). *Play-by-Play: Radio, television, and big-time college sport.* Baltimore, MD: Johns Hopkins University Press.

Stern, R. (1983). *They Were Number One: A history of the NCAA Basketball Tournament.* New York, NY: Leisure Press.

Thorn, J., Palmer, P., & Gershman, M. (1994). *Total Baseball* (4th ed.) New York, NY: Viking.

U.S. House of Representatives, Committee on Energy and Commerce, Subcommittee on Commerce, Trade, and Consumer Protection. (December 7, 2005). *Determining a champion on the field: A comprehensive review of the BCS and postseason college football.* Washington.

Wetzel, D. (January 9, 2002). RPI continues to lack an ounce of reason. *CBS Sportsline.*

Wetzel, D., Peter, J., & Passan, J. (2010). *Death to the BCS: The definitive case against the Bowl Championship Series.* New York, NY: Gotham Books.

CHAPTER 2

Regression Models

2.1 INTRODUCTION

In this chapter we provide readers with the math, probability, and statistics necessary to perform linear regression analysis. We show how to devise proper regression models to rank sports teams, predict the winning team and score, and calculate the probability of winning a contest and/or beat a sports betting line.

The models in this chapter are formulated to allow any sports fan or sports professional (including managers, coaches, general managers, agents, and weekend fantasy sports competitors) to easily apply to predict game outcomes, scores, probability of winning, and rankings using spreadsheet models. These techniques do not require an in-depth knowledge of statistics or optimization, and an advanced degree is not required.

Mathematical regression analysis includes three different types of regressions: linear regression analysis, log-linear regression analysis, and nonlinear regression analysis. In this chapter we will focus on linear regression analysis. Log-linear and nonlinear regression analyses are also appropriate techniques in many situations but are beyond the scope of this book.

In Chapter 3, Probability Models, we describe the different families of probability models to determine the probability of winning a game or achieving a specified outcome (such as winning margin or total points). In Chapter 4, Advanced Math and Statistics, we introduce advanced models and mathematical techniques that can be used to evaluate players and determine their overall contribution to winning, to determine the optimal mix of players to provide the best chances of winning based on the opponent and/or opponents' lineup, for picking fantasy sports teams, and for salary negotiations between team owners and agents. These models include principal component analysis and neural networks.

2.2 MATHEMATICAL MODELS

The usage of mathematical models and statistics in any professional application is to serve four main purposes:

1. Determine explanatory factors;
2. Determine the sensitivity of the dependent variable to explanatory factors;
3. Estimate outcome values;
4. Perform sensitivity and what-if analysis, to help determine how the outcome is expected to change if there is an unexpected change in the set of explanatory variables.

These are described as follows:

Determine Relationship Between Explanatory Factors and Dependent Variable. Mathematical models are employed by analysts to help determine a relationship between a dependent variable and a set of factors. The dependent variable, denoted as the y-variable, is the value that we are looking to determine and is often referred to as the outcome. The explanatory factors, denoted as the x-variables, are also often referred to as the independent factors, the predictor variables, or simply the model factors. These x-variables consist of the information set that we are using to help us gain a better understanding of the expected outcome (the y-variable). In sports models, this step will assist analyses, coaches, and managers in determining the most important set of factors to help predict the team's winning chances. For example, is it best to use batting average, on-base percentage, or slugging percentage to predict the number of runs a baseball team will score, or is it best to use some weighted combination of the three. Additionally, is it best to evaluate a pitcher based on earned run average (ERA), the opposing team's batting averages against the pitcher, the pitcher's strikeout-to-walk ratio, the opposing team's on-base percentage, or some combination of these statistics. And if it is best to use a combination of these measures, what is the best weighting to maximize likelihood of success.

Determine the Sensitivity of the Dependent Variable to These Factors. Mathematical techniques are employed to determine the sensitivity of the dependent variable to the set of explanatory factors. That is, how much is the dependent variable expected to change given different factors. After determining a statistically significant set of explanatory variables we then estimate the sensitivity of the dependent variable to each of these factors. This step will assist us in determining the best mix of factors to use to predict game outcomes, as well as the proper weighting across a group of entirely different or similar factors. The dependent variable that we are trying to predict can consist of many different outcomes such as the expected winning team, the probability that a team will win a game, the final score, and/or the expected winning margin or point spread.

Estimate Outcome Values. Mathematical models are used to estimate the outcome variable from the set of explanatory factors and sensitivities. The relationship will help determine the winning team, expected winning score, and the probability that the team will win. These models also help managers and coaches determine the best strategy to face an opponent based on different sets of factors and scenarios.

Perform Sensitivity and What-If Analysis. Mathematical models are used to help determine how the outcome is expected to change if there is an unexpected change to the explanatory factors. This will help coaches, managers, and analysts determine how the outcome of a game will change if a star player is not able to participate due to injury or suspension, if a team has acquired a new player through a trade or signing, as well as if a player is coming off an injury and is not at full strength. Furthermore, sensitivity and what-if analysis will help teams determine the best mix of players to use in a game based on the opponent or opposing team's lineup.

Statistics have long been used to describe the result of a match or game as well as to evaluate player or team performance. But in many cases, these models and statistics are not being applied correctly. Mathematical models are just now starting to be employed through all levels of sports to train players, help teams improve their chances of winning, and to determine and negotiate salaries. The sections below provide proper statistical methods and approaches for use in sports models and management.

2.3 LINEAR REGRESSION

Regression analysis is used to help us uncover a relationship between the dependent y-variable (what we are trying to predict) and the set of x-explanatory factors (which is input data that we are relying on to help us predict the outcome). Regression analysis will also help us determine if the explanatory factor is a statistically significant predictor of the outcome; if it is, it will help us calculation the sensitivity of the dependent variable to the explanatory factor.

A linear regression model is a model where we formulate a linear relationship between a dependent variable y and a set of explanatory factors denoted as x. In the event that there is more than one explanatory factor these factors are denoted as $x_1, x_2, \ldots, x_k$.

In the case where we have a single explanatory factor, the analysis is called a simple regression model and is written as:

$$y = b_0 + b_1 x + u$$

Here, y is the dependent variable or outcome (i.e., what we are looking to predict), x is the explanatory factor (i.e., what we are using to predict the outcome), and u is the regression error (the value of y that is not explained by the relationship with x). The error term u is also known as the noise of the model and signifies the quantity of y that was not explained by its factors.

In the above equation, b_0 and b_1 are the model parameters and define the sensitivity of the dependent variable and the explanatory factor. That is, how much will the dependent variable change based on the explanatory factor. In practice, however, the exact parameter values are not known and therefore must be estimated from the data.

The corresponding simple regression equation used to estimate the model parameters is:

$$\hat{y} = \hat{b}_0 + \hat{b}_1 x + \varepsilon$$

where $\hat{y}$ is the estimated dependent variable value, $\hat{b}_0$ is the constant term and $\hat{b}_1$ are the estimated parameters indicating the sensitivity of y to explanatory factor x, and ε is the regression error term, i.e., the value of y that is not explained by the regression model.

In the case where we have more than one explanatory factor, the analysis is called a multiple regression model and has the form:

$$y = b_0 + b_1 x_1 + b_2 x_2 + \cdots b_k x_k + u$$

Here, y is the dependent variable or outcome (what we are looking to predict), $x_1, x_2, \ldots, x_k$ are the k-explanatory factors, and u is the random noise (i.e., the value of y that is not explained by the set of explanatory factors).

Here b_0 represents the model constant and $b_1, \ldots, b_k$ represent the actual parameter values of the model. In practice, these exact values are not known in advance or with any certainty and must be determined from the data.

The corresponding multiple regression model used to estimate the outcome event y is:

$$\hat{y} = \hat{b}_0 + \hat{b}_1 x_1 + \hat{b}_2 x_2 + \cdots + \hat{b}_k x_k + \varepsilon$$

where $\hat{y}$ is the estimated dependent variable value, $\hat{b}_0$ is the model constant and $\hat{b}_1, \hat{b}_2, \ldots, \hat{b}_k$ are the estimated parameter values and model sensitivities to the factors, and ε is the regression noise. The goal of regression analysis is to help us determine which factors are true predictors of the explanatory variable y and the underlying sensitivity of the dependent variable y to the explanatory variables x.

Estimating Parameters

To estimate the parameters of the regression model we will first need to collect a set of observations. The more data samples we collect or have available the more accurate the prediction will be. A regression model with n observations can be written in vector form as follows:

$$y = b_0 + b_1 x_1 + b_2 x_2 + \cdots b_k x_k + u$$

where

$$
y = \begin{pmatrix} y_1 \\ y_2 \\ \vdots \\ y_t \\ \vdots \\ y_n \end{pmatrix} \quad
x_1 = \begin{pmatrix} x_{11} \\ x_{12} \\ \vdots \\ x_{1t} \\ \vdots \\ x_{1n} \end{pmatrix} \quad
x_2 = \begin{pmatrix} x_{21} \\ x_{22} \\ \vdots \\ x_{2t} \\ \vdots \\ x_{2n} \end{pmatrix} \quad
x_k = \begin{pmatrix} x_{k1} \\ x_{k2} \\ \vdots \\ x_{kt} \\ \vdots \\ x_{kn} \end{pmatrix} \quad
u = \begin{pmatrix} u_1 \\ u_2 \\ \vdots \\ u_t \\ \vdots \\ u_n \end{pmatrix}
$$

In this notation,

y_t = the value of the dependent variable at time t
x_{kt} = the value of the kth factor at time t
u_t = the value of the error term at time t

The expected y-value from our regression equation is:

$$\hat{y} = \hat{b}_0 + \hat{b}_1 x_1 + \hat{b}_2 x_2 + \cdots + \hat{b}_k x_k$$

The parameters are then estimated via a least-squares minimization process referred to as ordinary least squares (OLS). This process is as follows:

Step 1: Calculate the error between the actual dependent variable value y and the regression output estimate $\hat{y}$ for every time period t. This is:

$$e_t = y_t - \hat{y}_t$$

Step 2: Square the error term as follows:

$$e_t^2 = \left(y_t - \hat{y}_t\right)^2$$

Step 3: Substitute $\hat{y}_t = \hat{b}_0 + \hat{b}_1 x_{1t} + \hat{b}_2 x_{2t} + \cdots + \hat{b}_k x_{kt}$ as follows:

$$e_t^2 = \left(y_t - \left(\hat{b}_0 + \hat{b}_1 x_{1t} + \hat{b}_2 x_{2t} + \cdots + \hat{b}_k x_{kt}\right)\right)^2$$

Step 4: Define a loss function L to be the sum of the square errors as follows:

$$L = \sum_{t=0}^{n} \left(y_t - \left(\hat{b}_0 + \hat{b}_1 x_{1t} + \hat{b}_2 x_{2t} + \cdots + \hat{b}_k x_{kt}\right)\right)^2$$

Step 5: Estimate model parameter via minimizing the loss function. This is accomplished by differentiating L for each variable x and setting the equation equal to zero. This is:

$$\frac{dL}{dx_0} = -2\hat{b}_0 \cdot \sum_{t=1}^{n} \left(y_t - \hat{b}_0 x_{0t} - \hat{b}_1 x_{1t} - \hat{b}_2 x_{2t} - \cdots - \hat{b}_k x_{kt}\right) = 0$$

$$\frac{dL}{dx_k} = -2\hat{b}_1 \cdot \sum_{t=1}^{n} \left(y_t - \hat{b}_0 x_{0t} - \hat{b}_1 x_{1t} - \hat{b}_2 x_{2t} - \cdots - \hat{b}_k x_{kt}\right) = 0$$

$$\vdots$$

$$\frac{dL}{dx_k} = -2\hat{b}_k \cdot \sum_{t=1}^{n} \left(y_t - \hat{b}_0 x_{0t} - \hat{b}_1 x_{1t} - \hat{b}_2 x_{2t} - \cdots - \hat{b}_k x_{kt}\right) = 0$$

The parameters are then calculated by solving the system of $k+1$ equations and $k+1$ unknowns. Readers familiar with linear algebra and/or Gaussian elimination will without question realize that the above set of equations can only be solved if and only if the x vectors are all independent. Otherwise, we cannot determine a unique solution for the regression equation and model parameters.

The formulas for our important linear regression models statistics are provided below.

Estimated Parameters

The parameters of the linear regression model are found by solving the set of simultaneous equations above. These results are:

$$\hat{b}_0 = \hat{b}_1 \bar{x}_1 + \hat{b}_2 \bar{x}_2 + \cdots + \hat{b}_k \bar{x}_k$$

$$\hat{b}_1 = \frac{\sum_{i=1}^{n} (x_{1i} - \bar{x}_1)(y_i - \bar{y})}{\sum_{i=1}^{n} (x_{1i} - \bar{x}_1)^2}$$

$$\vdots$$

$$\hat{b}_k = \frac{\sum_{i=1}^{n} (x_{ki} - \bar{x}_k)(y_i - \bar{y})}{\sum_{i=1}^{n} (x_{ki} - \bar{x}_k)^2}$$

Se(b_k): Standard Error of the Parameter

Computing the standard error of the parameter is slightly more detailed than estimating the parameter values and can best be described in general terms using matrix algebra (unfortunately there is no way around this).

For a simply linear regression model and a model with two factors this calculation is a little more straightforward, but below we present the results for the general case of a k-factor regression model.

$$Cov(\hat{b}) = \sigma_y^2 \cdot (X'X)^{-1} = \begin{pmatrix} var(\hat{b}_0) & cov(\hat{b}_0, \hat{b}_1) & \cdots & cov(\hat{b}_0, \hat{b}_k) \\ cov(\hat{b}_1, \hat{b}_0) & var(\hat{b}_1) & \cdots & cov(\hat{b}_1, \hat{b}_k) \\ \vdots & \vdots & \ddots & \vdots \\ cov(\hat{b}_k, \hat{b}_0) & cov(\hat{b}_k, \hat{b}_1) & \cdots & var(\hat{b}_k) \end{pmatrix}$$

In this notation, the X variable denotes a data matrix with the first column being all 1s. This is now consistent with the expanded linear regression model above.

$$X = \begin{bmatrix} 1 & x_{11} & x_{21} & \cdots & x_{k1} \\ 1 & x_{12} & x_{22} & \cdots & x_{k2} \\ \vdots & \vdots & \vdots & \ddots & \vdots \\ 1 & x_{1n} & x_{2n} & \cdots & x_{kn} \end{bmatrix}$$

The variable X' denotes the transpose of the X matrix, and $(X'X)^{-1}$ represents the inverse of the product matrix $X'X$.

The term σ_y^2 denotes the regression variance, which is also known as the squared regression error value.

Then we have the variance of the parameter estimates as follows:

$$Var(\hat{b}) = \begin{pmatrix} var(\hat{b}_0) \\ var(\hat{b}_1) \\ \vdots \\ var(\hat{b}_k) \end{pmatrix}$$

And finally, the standard error of the parameter estimates is:

$$Se(\hat{b}) = \begin{pmatrix} \sqrt{var(\hat{b}_0)} \\ \sqrt{var(\hat{b}_1)} \\ \vdots \\ \sqrt{var(\hat{b}_k)} \end{pmatrix}$$

Regression Error

$$\sigma_y^2 = \frac{\sum e_i^2}{n - k}$$

R-Square (R^2)

$$R^2 = 1 - \frac{\sum e_i^2}{\sum y_i^2}$$

t-Stat

$$t_k = \frac{\hat{b}_k}{Se(\hat{b}_k)}$$

F-Value

$$F = \frac{R^2/(k-1)}{(1 - R^2)(n - k)}$$

Regression Model Requirements

A proper regression model and analysis needs to ensure that the regression model satisfies the required model properties. Here, there are seven main properties of a linear regression model. If any of these assumptions are violated, the results of the analysis could be suspect and potentially give incorrect insight into the true relationship between dependent variable and factors.

In these cases, analysts need to make adjustments to the data. These techniques are further explained in Gujarati (1988), Kennedy (1998), and Greene (2000).

The main assumptions of the linear regression model are:

A1. Linear relationship between dependent variable and explanatory factors.

$$y = b_0 + b_1 x_1 + \cdots + b_k x_k + u$$

A2. Unbiased parameters: the estimated parameter values are unbiased estimates of the turn parameter values. That is, they satisfy the following relationship:

$$E(b_0) = b_0, E(b_1) = b_1, \ldots, E(b_k) = b_k$$

A3. Error term mean zero: the expected value of the error term is zero.

$$E(\varepsilon) = 0$$

A4. Constant variance: each error term has the same variance (i.e., no heteroskedasticity).

$$Var(\varepsilon_k) = \sigma^2 \text{ for all } k$$

A5. Independent error t term: no autocorrelation or correlation of any degree.

$$E(\varepsilon_k \varepsilon_{k-t}) = 0 \text{ for all lagged time periods } t$$

A6. Errors are independent of explanatory factors.

$$Cov(\varepsilon, x_k) = 0 \text{ for all factors } k$$

A7. Explanatory factors are independent.

$$Cov(x_j, x_k) = 0 \text{ for all factors } j \text{ and } k$$

2.4 REGRESSION METRICS

In performing regression analysis and evaluating the model, we need the following set of statistical metrics and calculations:

$\hat{b}_k$ = model parameter values (estimated sensitivity of y to factor k)
e = regression error (determined from the estimation process)
$Se(b_k)$ = standard error of the estimated parameter b_k
σ_y = standard error of the regression model
R^2 = goodness of fit (the percentage of overall variance explained by the model)
t-Stat = critical value for the estimated parameter
F-Stat = critical value for the entire model

Regression Analysis Statistics

In this section we provide readers with an overview of the important statistics for performing regression analysis. Our summary is not intended to be a complete listing of the required analysis and evaluate metrics, but these will serve as an appropriate starting place for the analyses. These important statistics are:

t-Test

The t-test is used to test the null hypothesis that a parameter value is zero. This would indicate that the selected explanatory factor does not have any predictive power in our regression equation.

The corresponding t-Stat for parameter k is:

$$t\text{-}stat(k) = \hat{\beta}(k)/Se\left(\hat{\beta}\right)$$

where

$\hat{\beta}(k)$ = parameter k

$Se\left(\hat{\beta}\right)$ = standard error of parameter k

$t\text{-}Stat(k)$ = t-statistic for parameter k

The testing hypothesis is:

$$H_0: \beta_k = 0$$
$$H_1: \beta_k \neq 0$$

Analysts could also test the alternative hypothesis that the parameter value is greater than or less than zero depending on the goal of the analysis. A general rule of thumb is that if the absolute value of the t-Stat above is greater than two, then reject the null hypothesis and conclude that factor "k" is a significant predictor variable.

That is:

if $|t_k| > 2$ then reject the null

R² Goodness of Fit

The R^2 statistic is a measure of the goodness of fit of a regression model. This statistic is also known as the coefficient of determinant. A regression model with strong explanatory power will have a high coefficient of variation R^2.

$$R^2 = 1 - \frac{Residual\ Sum\ of\ Squares}{Total\ Sum\ of\ Squares}$$

F-Test

The F-test is used in regression analysis to test the hypothesis that all model parameters are zero. It is also used in statistical analysis when comparing statistical models that have been fitted using the same underlying factors and data set to determine the model with the best fit. That is:

$$H_0: B_1 = B_2 = \cdots = B_k = 0$$

$$H_1: B_j \neq 0 \text{ for at least one } j$$

The F-test was developed by Ronald A. Fisher (hence F-test) and is a measure of the ratio of variances. The F-statistic is defined as:

$$F = \frac{Explained\ variance}{Unexplained\ variance}$$

A general rule of thumb that is often used in regression analysis is that if $F > 2.5$ then we can reject the null hypothesis. We would conclude that there is a least one parameter value that is nonzero.

Probability Home Team Wins

The probability that the actual outcome Y-variable will be less than or equal to a value S can be computed from the estimated Y-variable and regression error as follows:

$$Prob(Y \leq S) = NormCdf(S, \hat{Y}, \sigma_y)$$

Here, $NormCdf(S, \hat{Y}, \sigma_y)$ represents the cumulative normal distribution with mean $= \hat{Y}$ and standard deviation $= \sigma_y$, and is the probability corresponding to a value less than or equal to S (see chapter: Advanced Math and Statistics, for detailed description of the calculation).

Subsequently, the probability that the home team will win a game if the expected outcome is $\hat{Y}$ and the regression standard error is σ_y is calculated as follows:

$$Prob(Home\ Team\ Wins) = 1 - NormCdf(0, \hat{Y}, \sigma_y)$$

This equation denotes the probability that the home team victory margin will be greater than zero and thus the home team wins the game.

Matrix Algebra Techniques

In matrix notation, the full regression model is written as:

$$y = X\beta + \varepsilon$$

The model used for estimation is:

$$\hat{y} = X\hat{\beta}$$

The vector of error terms (also known as vector of residuals) is then:

$$e = y - X\hat{\beta}$$

Estimate Parameters

The parameters of our regression model are estimated via OLS as follows:

Step I: Compute the residual sum of squares:

$$e^T e = \left(y - X\hat{\beta}\right)^T \left(y - X\hat{\beta}\right)$$

Step II: Estimate the parameters $\hat{\beta}$ via differentiating. This yields:

$$\hat{\beta} = \left(X^T X\right)^{-1} X^T y$$

Compute Standard Errors of $\hat{\beta}$

This is calculated by computing the covariance matrix of $\hat{\beta}$. We follow the approach from Greene (2000) and Mittelhammer, Judge, and Miller (2000). This is as follows:

Step I: Start with the estimated $\hat{\beta}$ from above and substitute for y.

$$
\begin{aligned}
\hat{\beta} &= \left(X^T X\right)^{-1} X^T y \\
&= \left(X^T X\right)^{-1} X^T (X\beta + \varepsilon) \\
&= \left(X^T X\right)^{-1} X^T X\beta + \left(X^T X\right)^{-1} X^T \varepsilon \\
&= I\beta + \left(X^T X\right)^{-1} X^T \varepsilon \\
&= \beta + \left(X^T X\right)^{-1} X^T \varepsilon
\end{aligned}
$$

Therefore, our estimated parameters are:

$$\hat{\beta} = \beta + \left(X^T X\right)^{-1} X^T \varepsilon$$

Step II: Computed expected $\hat{\beta}$ as follows:

$$E\left(\hat{\beta}\right) = E\left(\beta + \left(X^T X\right)^{-1} X^T \varepsilon\right)$$
$$= E(\beta) + E\left(\left(X^T X\right)^{-1} X^T \varepsilon\right)$$
$$= E(\beta) + \left(X^T X\right)^{-1} X^T E(\varepsilon)$$
$$= \beta + \left(X^T X\right)^{-1} X^T \cdot 0$$
$$= \beta$$

Therefore, we have

$$E\left(\hat{\beta}\right) = \beta$$

which states $\hat{\beta}$ is an unbiased estimate of β

Step III: Compute the covariance matrix of $\hat{\beta}$ as follows:

$$Cov\left(\hat{\beta}\right) = E\left(\left(\hat{\beta} - \beta\right)\left(\hat{\beta} - \beta\right)^T\right)$$
$$= E\left(\left(\left(X^T X\right)^{-1} X^T \varepsilon\right)\left(\left(X^T X\right)^{-1} X^T \varepsilon\right)^T\right)$$
$$= E\left(\left(X^T X\right)^{-1} X^T \varepsilon \varepsilon^T X \left(X^T X\right)^{-1}\right)$$
$$= \left(X^T X\right)^{-1} X^T E\left(\varepsilon \varepsilon^T\right) X \left(X^T X\right)^{-1}$$
$$= \left(X^T X\right)^{-1} X^T \left(\sigma^2 \cdot I\right) X \left(X^T X\right)^{-1}$$
$$= \sigma^2 \cdot \left(X^T X\right)^{-1} X^T X \left(X^T X\right)^{-1}$$
$$= \sigma^2 \cdot I \left(X^T X\right)^{-1}$$
$$= \sigma^2 \left(X^T X\right)^{-1}$$

It is important to note that if $E\left(\varepsilon \varepsilon^T\right) \neq \sigma^2 \cdot I$ then the data is heteroskedastic, i.e., it is not constant variance and it violates one of our required regression properties.

The standard error of the parameters is computed from the above matrix:

$$Se\left(\hat{\beta}\right) = diag\left(\sqrt{\sigma^2 (X^T X)^{-1}}\right)$$

R^2 Statistic

$$R^2 = \frac{\hat{b}' X' y - n\bar{y}^2}{y' y - n\bar{y}^2}$$

The coefficient of determination will be between 0 and 1. The closer the value is to 1, the better the fit of the model.

F-Statistic

$$F = \frac{\left(\hat{b}' X' y - n\bar{y}^2\right)/(k-1)}{\left(y'y - \hat{b}' X' y\right)/(n-k)}$$

Probability Home Team Wins

The probability that the actual outcome Y-variable will be less than or equal to a value S can be computed from the estimated Y-variable and regression error as follows:

$$Prob(Y \leq S) = NormCdf(S, \hat{Y}, \sigma_y)$$

The probability that the home team will win a game if the expected outcome is $\hat{Y}$ and the regression standard error is σ_y is:

$$Prob(Home\ Team\ Wins) = 1 - NormCdf(0, \hat{Y}, \sigma_y)$$

(See chapter: Advanced Math and Statistics, for detailed explanation.)

Example 2.1 R^2 Goodness of Fit

In this section we provide the reader with different examples of the application of linear regression models and detailed discussion on how to interpret the regression results. These results are as follows.

Fig. 2.1 provides an illustration of four different linear regression models of the form $y = b_0 + b_1 x + e$ with different R^2 metrics and goodness of fit. In these models, x represents the input value and y represents is the predicted value.

Fig. 2.1A is an example of a model with no relationship between the input variable and predicted y-value. This model has a very low goodness of fit $R^2 = 0.02$ and has a large amount of scatter and variation in the results with no relationship between y and x. Fig. 2.1B is an example of a model with some predictive power and a goodness of fit $R^2 = 0.22$. Notice the relationship between x and y in this example, but with some scatter in the results. Fig. 2.1C is an example of a strong predictive model and high goodness of fit $R^2 = 0.57$. Notice the relationship between the y-value and input data x and a distinguishable trend line. Fig. 2.1D is an example of a model with very strong predictive power and very high goodness of fit $R^2 = 0.85$. Notice in this example that most of the predicted data lies very near or on the trend line.

An important question that often arises regarding the R^2 goodness of fit metric is what is the appropriate value to use as selection criteria for the acceptance or rejection of a model or data set. This is a difficult question to answer because it is dependent upon the overall variation or noise in the data. Some statistical relationships can be predicted with a great degree of accuracy because we have all the important explanatory

R^2 Goodness of Fit

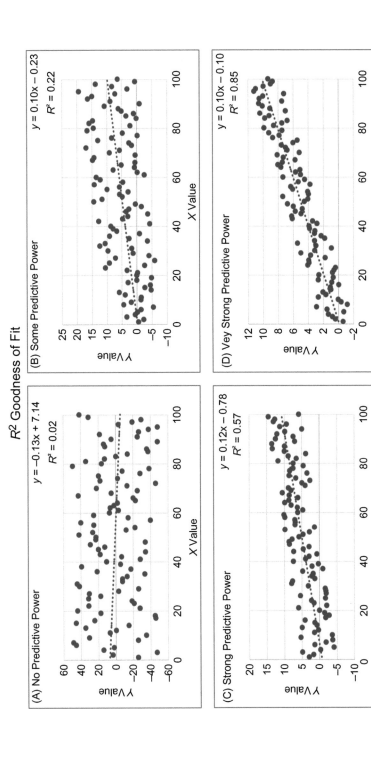

Figure 2.1 R^2 Goodness of Fit: (A) No Predictive Power, (B) Some Predictive Power, (C) Strong Predictive Power, (D) Very Strong Predictive Power.

Table 2.1 Team Rating Table	
Team	Rating
1	2.141
2	1.361
3	4.563
4	1.148
5	2.524
6	3.106
7	4.877
8	6.064
9	2.617
10	5.752

factors and there is relatively little noise or uncertainty in the output results. Other models may have significant dependency on some set of explanatory input factors but also have a large amount of variation in the output data. In these cases, the best fit model may only have a small R^2.

As we show in subsequent chapters, a good fit sports prediction model will have an R^2 goodness of fit metric from $R^2 = 0.20$ through $R^2 = 0.40$. Values higher ($R^2 > 0.40$) are considered excellent models and lower values ($R^2 < 0.20$) are considered less predictable.

It is important to note that the R^2 goodness of fit should not be the sole metric used to critique a linear regression model or select the appropriate underlying data set, but it does provide insight into the model's predictive power and overall goodness of fit.

Linear Regression Example

We next provide three examples showing a process to evaluate a linear regression sports prediction model. In these examples, the output variable Y is the home team's spread (i.e., the home team points minus the away team points); this will be a common theme throughout subsequent chapters. The input data that will be used to predict the home team's spread consists of game results data including team strength rating (Rating), game attendance (Attend), and game time temperature (Degrees).

The data for these models and observations is provided in Tables 2.1 and 2.2. The general form for this model is:

$$Y = b_0 + b_1 \cdot X_H - b_2 \cdot X_A + b_3 \cdot Attend + b_4 \cdot Degrees$$

Example 2.2 Regression: Team Rating

In this example, the linear regression prediction model will only incorporate the team's strength rating to make a prediction. This regression was performed in Excel using the = LINEST() function and the results are shown in Table 2.3.

These results show a very high $R^2 = 0.75$ and all the variables are significant with $|t| > 2$. The signs of the beta estimates are intuitive. The higher the home team strength rating the higher the spread but the higher the away team strength rating the lower the spread. Thus, if you are the home team and play a weaker team you are expected to win, and if you play a stronger team you are expected to win by fewer points or possibly lose.

Another important item in this regression is the intercept term $b_0 = 2.87$. This signifies that there is a home field advantage of 2.87 points; thus, a benefit of playing at home.

For example, there may be certain benefits associated with playing at home due to the field dimensions (such as baseball), field and wind (such as football and soccer), or process (such as baseball with the home team batting last) that truly provide an advantage to the home team. Another possibility is that there is a benefit from playing in familiar surroundings and proper rest, as opposed to being worn out from a busy sports travel schedule.

Example 2.3 Regression: Team Rating and Attendance

This example incorporates game attendance into the regression along with team strength rating. The game attendance value was computed as the actual game attendance minus the average league attendance over the season and is expressed in thousands. Therefore, if the average attendance was 25,000 fans and the actual game had 35,500 fans the attendance variable is 15.5. If the game attendance was 20,500 fans the attendance variable is -4.5.

This regression was performed in Excel using the = LINEST() function and the results are shown in Table 2.4.

These results of this model again have a very high $R^2 = 0.78$ and all the variables are again significant with $|t| > 2$. The signs of the beta estimates are also intuitive and the team rating parameters are similar to the previous example. Once again we have a home field advantage since $b_0 = 2.71$ ($t = 2.97 > 2$), and also the attendance parameter is positive $b_3 = 0.13$ with a significant t-test ($t = 3.26 > 2$).

The interesting finding here is that there could be two different home field advantage items occurring. First, we have the familiarity of the field or layout (if applicable since some sports have the exact same dimensions and are played indoors) or scoring method (e.g., in baseball the home team bats last and has a true benefit) or just more rest and familiarity with surroundings. However, there is also a benefit driven by the number of fans at the game, which will usually provide the home team with extra encouragement and perhaps a psychological edge.

Our regression can provide insight into many different potential cause—effect relationships, many of which would be extremely difficult to ascertain with the use of mathematics.

Example 2.4 Regression: Team Rating, Attendance, and Temperature

In our last example we incorporate the day's temperature in degrees into the model to determine if this has any effect on the outcome. It is an important variable to incorporate for some sports such as football, where there could be large difference in temperature for

Table 2.2 Game Results

Game Number	Home Team	Away Team	Home Rating	Away Rating	Attendance (Thousands)	Temperature (Degrees)	(Home–Away) Point Difference
1	1	2	2.141	1.361	−3.20	72.63	6
2	1	3	2.141	4.563	11.59	95.43	1
3	1	4	2.141	1.148	9.36	55.52	7
4	1	5	2.141	2.524	−11.84	55.81	−1
5	1	6	2.141	3.106	10.39	95.75	1
6	1	7	2.141	4.877	9.00	66.28	−3
7	1	8	2.141	6.064	−6.02	94.33	1
8	1	9	2.141	2.617	−1.23	59.59	−1
9	1	10	2.141	5.752	−9.35	63.27	−7
10	2	1	1.361	2.141	1.21	62.06	−1
11	2	3	1.361	4.563	0.44	65.46	−6
12	2	4	1.361	1.148	−5.93	97.24	3
13	2	5	1.361	2.524	0.87	59.28	−1
14	2	6	1.361	3.106	10.53	74.06	−3
15	2	7	1.361	4.877	−10.07	82.05	−1
16	2	8	1.361	6.064	−4.10	90.79	−6
17	2	9	1.361	2.617	−4.99	79.48	−3
18	2	10	1.361	5.752	−9.69	63.64	−4
19	3	1	4.563	2.141	−1.76	59.70	10
20	3	2	4.563	1.361	2.19	75.69	7
21	3	4	4.563	1.148	−5.16	58.60	10
22	3	5	4.563	2.524	−8.81	75.72	7
23	3	6	4.563	3.106	−0.65	98.85	7
24	3	7	4.563	4.877	5.16	92.08	6
25	3	8	4.563	6.064	−9.70	86.23	−1
26	3	9	4.563	2.617	−3.17	52.09	9
27	3	10	4.563	5.752	−2.17	53.91	−1
28	4	1	1.148	2.141	−2.08	73.67	−2
29	4	2	1.148	1.361	8.24	93.19	7
30	4	3	1.148	4.563	3.88	76.82	−2
31	4	5	1.148	2.524	7.25	94.85	−2
32	4	6	1.148	3.106	−9.29	92.13	1
33	4	7	1.148	4.877	8.59	62.32	−7
34	4	8	1.148	6.064	0.91	79.25	−13
35	4	9	1.148	2.617	−6.45	58.96	−4
36	4	10	1.148	5.752	−11.25	62.62	−11
37	5	1	2.524	2.141	−2.98	85.99	9
38	5	2	2.524	1.361	11.83	63.63	9
39	5	3	2.524	4.563	−9.42	96.64	−4
40	5	4	2.524	1.148	−5.37	70.98	10
41	5	6	2.524	3.106	5.52	56.46	6
42	5	7	2.524	4.877	4.23	60.89	3
43	5	8	2.524	6.064	4.11	61.90	−4
44	5	9	2.524	2.617	−11.90	79.27	−6
45	5	10	2.524	5.752	4.52	96.16	−5
46	6	1	3.106	2.141	9.47	81.36	5
47	6	2	3.106	1.361	−9.88	75.27	5
48	6	3	3.106	4.563	−1.08	78.81	5
49	6	4	3.106	1.148	−1.19	72.10	10
50	6	5	3.106	2.524	5.81	95.25	4
51	6	7	3.106	4.877	−3.87	70.73	1
52	6	8	3.106	6.064	6.90	67.32	−3
53	6	9	3.106	2.617	9.47	67.37	9
54	6	10	3.106	5.752	−8.61	94.29	−3

(*Continued*)

Table 2.2 (Continued)

Game Number	Home Team	Away Team	Home Rating	Away Rating	Attendance (Thousands)	Temperature (Degrees)	(Home−Away) Point Difference
55	7	1	4.877	2.141	−9.31	71.86	6
56	7	2	4.877	1.361	11.44	98.53	9
57	7	3	4.877	4.563	−5.69	66.22	2
58	7	4	4.877	1.148	−7.57	64.85	12
59	7	5	4.877	2.524	7.33	54.28	11
60	7	6	4.877	3.106	6.99	63.19	10
61	7	8	4.877	6.064	−7.90	64.10	1
62	7	9	4.877	2.617	10.30	64.33	9
63	7	10	4.877	5.752	12.09	52.14	−1
64	8	1	6.064	2.141	2.80	79.05	10
65	8	2	6.064	1.361	−5.86	88.39	10
66	8	3	6.064	4.563	−4.44	76.51	4
67	8	4	6.064	1.148	−10.02	64.00	15
68	8	5	6.064	2.524	9.44	75.53	17
69	8	6	6.064	3.106	−4.08	80.96	10
70	8	7	6.064	4.877	3.77	56.78	14
71	8	9	6.064	2.617	−9.99	50.07	10
72	8	10	6.064	5.752	−3.75	86.09	6
73	9	1	2.617	2.141	−10.45	76.95	6
74	9	2	2.617	1.361	2.78	81.96	5
75	9	3	2.617	4.563	−2.72	77.61	−2
76	9	4	2.617	1.148	7.46	69.47	10
77	9	5	2.617	2.524	3.40	99.78	−4
78	9	6	2.617	3.106	1.93	68.81	1
79	9	7	2.617	4.877	−2.74	93.60	−1
80	9	8	2.617	6.064	−9.78	64.17	−5
81	9	10	2.617	5.752	11.83	63.23	1
82	10	1	5.752	2.141	2.39	89.07	10
83	10	2	5.752	1.361	−9.47	56.04	13
84	10	3	5.752	4.563	−6.11	59.22	8
85	10	4	5.752	1.148	4.74	71.15	11
86	10	5	5.752	2.524	−7.94	67.31	11
87	10	6	5.752	3.106	9.33	63.25	4
88	10	7	5.752	4.877	−9.21	80.72	2
89	10	8	5.752	6.064	9.62	86.99	3
90	10	9	5.752	2.617	−7.89	63.79	5

Table 2.3 Team Rating Regression Results

	Intercept	Home Rating	Away Rating
Beta	2.41	2.27	−2.04
SE	1.05	0.20	0.20
t-Stat	2.29	11.57	−10.40
R^2	0.76		
SeY	3.12		
F	135.97		

Table 2.4 Team Rating and Attendance Regression Results

	Intercept	Home Rating	Away Rating	Attendance
Beta	2.27	2.30	−2.01	0.10
SE	1.03	0.19	0.19	0.04
t-Stat	2.22	11.99	−10.52	2.35
R^2	0.77			
SeY	3.04			
F	97.21			

Table 2.5 Team Rating, Attendance, and Temperature Regression Results

	Intercept	Home Rating	Away Rating	Attendance	Degrees
Beta	3.10	2.29	−2.01	0.10	−0.01
SE	2.11	0.19	0.19	0.04	0.02
t-Stat	1.47	11.74	−10.45	2.36	−0.45
R^2	0.77				
SeY	3.06				
F	72.28				

different teams' locations, which could provide them with benefits gained by training in these weather conditions. For example, consider the difference in outdoor temperature between Buffalo and Miami. The results for this regression are in Table 2.5; they show how to determine if an explanatory factor is not a significant predictor variable.

First, the model does have a high goodness of fit with $R^2 = 0.78$. But we do not have significant t-Stats for all our variables. The beta for the Degrees variable is $b_4 = 0.02$ but it does not have a significant t-Stat for temperature ($t = 1.02 > 2$), thus indicating that temperature is not a predictor of home margin.

In this case, it would not be appropriate to incorporate the temperature variable into the regression and this model formulation should not be used for sports predictions. We should revert back to the model in the previous examples.

Example 2.5 Estimating Home Team Winning Margin

This example shows how we can use the linear regression model to determine the likely winner of a game. Suppose that we are interested in predicting the outcome of a game between Team #6 playing at home against Team #7 in front of 20,000 fans. The estimation equation is based on Example 2.3 and has the form:

$$Y = b_0 + b_1 \cdot \text{Home Rating} - b_2 \cdot \text{Away Rating} + b_3 \cdot \text{Attendance}$$

From Table 2.1 we have Team #6 Rating $= 3.106$ and Team #7 Rating $= 4.877$. Using the regression results for Example 2.3 shown in Table 2.4 we have the following:

$$Est.\ Home\ Team\ Margin = 2.27 + 2.30 \cdot 3.106 - 2.01 \cdot 4.877 + 0.10 \cdot 20 = 1.66$$

Therefore, home team #6 is expected to win by 1.66 points.

The probability that the home team will win is based on the expected winning margin of 1.66 points and the regression standard error of 3.06. This is computed as follows:

$$Prob(Home\ Wins) = 1 - NormCdf(0, 1.66, 3.06) = 71\%$$

Thus, the home team has a 71% winning probability.

2.5 LOG-REGRESSION MODEL

A log-regression model is a regression equation where one or more of the variables are linearized via a log-transformation. Once linearized, the regression parameters can be estimated following the OLS techniques above. It allows us to transform a complex nonlinear relationship into a simpler linear model that can be easily evaluated using direct and standard techniques.

Log-regression models fall into four categories: (1) linear model, which is the traditional linear model without making any log transformations; (2) linear-log model, where we transform the x-explanatory variables using logs; (3) log-linear model, where we transform the y-dependent variable using logs; and (4) a log-log model, where both the y-dependent variable and the x-explanatory factors are transformed using logs.

For example, if Y and X refer to the actual data observations, then our four categories of log transformations are:

1. Linear: $Y = b_0 + b_1 \cdot X + u$
2. Linear-Log: $Y = b_0 + b_1 \cdot \log(X) + u$
3. Log-Linear: $\log(Y) = b_0 + b_1 \cdot X + u$
4. Log-Log: $\log(Y) = b_0 + b_1 \cdot \log(X) + u$

As stated, the parameters of these models can be estimated directly from our OLS technique provided above.

The process to determine the values for Y from the estimated $\log(Y)$ using log-regression categories (3) and (4) is not as direct and requires further input from the regression analysis error term. For example, if a variable y has a log-normal distribution with mean u and variance v^2, i.e.,

$$y \sim \log Normal(u, v^2)$$

Then the expected value of $E(y)$ is calculated as follows:

$$E(\log(y)) = u + \frac{1}{2} \cdot v^2$$

And, therefore, we have

$$y = e^{\left(u + \frac{1}{2} v^2\right)}$$

Notice that the variance term is included in the estimation equation.

If the estimate parameters of the log-linear equation are $\hat{b}_0$ and $\hat{b}_1$ and regression error variance term is σ^2, then we estimate our $\log(y)$ as follows:

$$E\left[\log(Y)\right] = \hat{b}_0 + \hat{b}_1 \cdot x + 0.5 \cdot \sigma^2$$

And Y is then determined as follows:

$$Y = e_0^{\hat{b} + \hat{b}_1 \cdot x + 0.5 \cdot \sigma^2}$$

Similarly, the estimated value of the Log-Log regression model is:

$$E\left[\log(Y)\right] = \hat{b}_0 + \hat{b}_1 \cdot \log(X) + 0.5 \cdot \sigma^2$$

which can be written in terms of Y as follows:

$$\hat{Y} = e^{\hat{b}_0 + \hat{b}_1 \cdot \log(X) + 0.5 \cdot \sigma^2}$$

Rewritten we have:

$$\hat{Y} = e^{\hat{b}_0 + 0.5 \cdot \sigma^2} e^{\hat{b}_1 \cdot \log(X)} = e^{\hat{b}_0 + 0.5 \cdot \sigma^2} X^{\hat{b}_1}$$

Now, if we let $\hat{k} = e^{\hat{b}_0 + 0.5 \cdot \sigma^2}$, then the regression estimate model can be written as:

$$\hat{Y} = k \cdot X_1^{\hat{b}}$$

let the regression equation be as follows:

$$Y = b_0 X_1^{b1} X_2^{b2} \varepsilon$$

If Y has log-normal distribution, then $\ln Y$ has a normal distribution. Thus

$$\ln(Y) = b_0 + b_1 X_1 + b_2 X_2 + \varepsilon$$

Example: Log-Transformation

An example of a regression model that can be solved through a log-transformation of the data is shown in Table 2.6 Log-Linear Regression Data. This model has form:

$$y = b_0 \cdot x^{b1} \varepsilon$$

Table 2.6 Log-Linear Regression Data

Obs	X	Y	Log(X)	Log(Y)
1	0.010	0.286	−4.605	−1.253
2	0.010	0.142	−4.605	−1.954
3	0.010	0.160	−4.605	−1.831
4	0.010	0.250	−4.605	−1.388
5	0.010	0.213	−4.605	−1.548
6	0.100	0.610	−2.303	−0.494
7	0.100	0.553	−2.303	−0.593
8	0.100	0.661	−2.303	−0.414
9	0.100	0.723	−2.303	−0.324
10	0.100	0.668	−2.303	−0.403
11	0.200	0.998	−1.609	−0.002
12	0.200	0.702	−1.609	−0.353
13	0.200	0.824	−1.609	−0.193
14	0.200	0.830	−1.609	−0.186
15	0.200	0.665	−1.609	−0.407
16	0.300	0.775	−1.204	−0.254
17	0.300	0.821	−1.204	−0.197
18	0.300	0.943	−1.204	−0.058
19	0.300	1.232	−1.204	0.209
20	0.300	1.120	−1.204	0.114
21	0.400	1.041	−0.916	0.040
22	0.400	1.108	−0.916	0.103
23	0.400	1.124	−0.916	0.117
24	0.400	1.376	−0.916	0.319
25	0.400	1.162	−0.916	0.150
26	0.500	1.521	−0.693	0.419
27	0.500	1.027	−0.693	0.027
28	0.500	1.354	−0.693	0.303
29	0.500	1.307	−0.693	0.267
30	0.500	1.528	−0.693	0.424
31	0.600	2.065	−0.511	0.725
32	0.600	2.047	−0.511	0.716
33	0.600	1.709	−0.511	0.536
34	0.600	1.527	−0.511	0.423
35	0.600	1.348	−0.511	0.299
36	0.700	1.900	−0.357	0.642
37	0.700	2.069	−0.357	0.727
38	0.700	1.982	−0.357	0.684
39	0.700	1.357	−0.357	0.305
40	0.700	1.902	−0.357	0.643
41	0.800	1.818	−0.223	0.598
42	0.800	1.241	−0.223	0.216
43	0.800	1.514	−0.223	0.415
44	0.800	1.650	−0.223	0.501
45	0.800	2.083	−0.223	0.734
46	0.900	1.668	−0.105	0.512
47	0.900	2.195	−0.105	0.786
48	0.900	2.319	−0.105	0.841
49	0.900	1.811	−0.105	0.594
50	0.900	2.112	−0.105	0.748
51	1.000	2.435	0.000	0.890
52	1.000	2.704	0.000	0.995
53	1.000	1.468	0.000	0.384
54	1.000	2.116	0.000	0.749
55	1.000	1.518	0.000	0.418

where $\ln(\varepsilon) \sim N(0, \sigma^2)$. Notice the nonlinear relationship between the dependent variable y and the explanatory variable x. The sensitivities b_0 and b_1 in this case can be determined via a log-transformation regression.

There is a linear relationship between the dependent variable Y and explanatory variable x. That is:

$$\ln(y) = \ln(b_0) + b_1\ln(x) + \varepsilon$$

The parameters of this model as determined via the OLS regression technique described above use the following formulation:

$$\ln(y) = \alpha_0 + \alpha_1\ln(x) + \varepsilon$$

Solving, we obtain:

$$\alpha_0 = 0.677$$
$$\alpha_1 = 0.503$$
$$\sigma_y = 0.190$$

The original parameters are finally computed as follows:

$$b_0 = \exp\left\{\alpha_0 + \sigma_y^2\right\} = \exp\left\{0.677 + 0.190^2\right\} = 2.003$$

$$b_1 = \alpha_1 = 0.503$$

And the best fit equation is:

$$y = 2.003 \cdot x^{0.503}\varepsilon$$

This relationship is shown in Fig. 2.2, where Fig. 2.2A shows the relationship between y and x for actual data and Fig. 2.2B shows the relationship between the log-transformed data. The regression results for the log-transformed data and parameters and the adjusted parameters are shown in Table 2.7.

2.6 NONLINEAR REGRESSION MODEL

Now let us turn our attention to nonlinear regression models. These models are comprised of nonlinear equations that cannot be linearized via a log-transformation. One of the more infamous nonlinear models is the I-Star market impact model introduced by Kissell and Malamut (1999) used for electronic, algorithmic, and high-frequency trading. See Kissell, Glantz, and Malamut (2004), Kissell and Malamut (2006), or Kissell (2013) for an overview of this model and its applications.

Log-Linear Regression Model

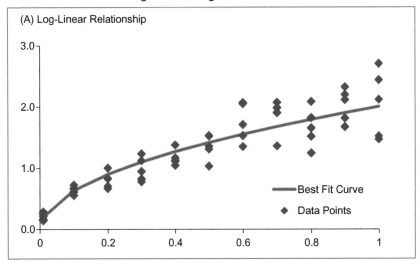

(A) Log-Linear Relationship

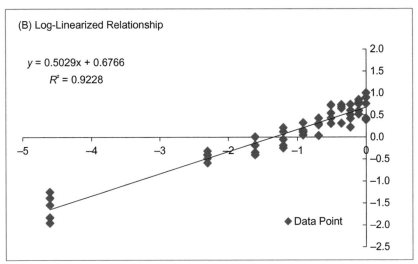

(B) Log-Linearized Relationship

$y = 0.5029x + 0.6766$
$R^2 = 0.9228$

Figure 2.2 Log-Linear Regression Model: (A) Log-Linear Relationship; (B) Log-Linearized Relationship.

The model has the form:

$$\hat{Y} = \hat{a}_0 \cdot X_1^{\hat{a}1} \cdot X_2^{\hat{a}2} \cdot X_3^{\hat{a}3} + \left(\hat{a}_5 X_4^{\hat{a}4} + (1 - \hat{a}_5)\right) + \varepsilon$$

where

Y = market impact cost of an order. This is the price movement in the stock due to the buying and selling pressure of the order or trade. It is comprised of a temporary and permanent component

Table 2.7 Log-Linear Regression Results

Category	Log-Adjustment		Actual Parameters	
	a_0	a_1	b_0	b_1
Beta	0.677	0.503	2.003	0.503
SE	0.034	0.020	0.034	0.020
t-Stat	19.764	25.178	58.506	25.178
R^2	0.923		0.923	
SeY	0.190		0.190	
F	633.946		633.946	

X_1 = order size as a percentage of average daily volume to trade (expressed as a decimal)

X_2 = annualized volatility (expressed as a decimal)

X_3 = asset price (expressed in local currency)

X_4 = percentage of volume and used to denote the underlying trading strategy (expressed as a decimal)

The parameters of the model are: $\hat{a}_0, \hat{a}_1, \hat{a}_2, \hat{a}_3, \hat{a}_4$, and $\hat{b}_1$.

The error term of the model ε has normal distribution with mean zero and variance v^2, i.e., $\varepsilon \sim N(0, v^2)$.

Nonlinear models of form can be solved via nonlinear OLS providing the error term has a normal distribution such as the famous I-Star model above. For nonlinear regression models where the underlying error distribution is not normally distributed we need to turn to maximum likelihood estimation techniques.

A process to estimate the parameters of nonlinear models is described in Greene (2000), Fox (2002), and Zhi, Melia, Guericiolini, et al. (1994).

We outline the parameter estimation process for a general nonlinear model below:

Step I: Define the model:

$$y = f(x, \beta) + \varepsilon$$

where

$$\varepsilon \sim iid(0, \sigma^2)$$

Let,

$$f(x; a_0, a_1, a_2, a_3, a_4, b_1) = a_0 \cdot X_1^{a1} \cdot X_2^{a2} \cdot X_3^{a3} + \left(b_1 \cdot X_4^{a4} + (1 - b_1)\right) + \varepsilon$$

Step II: Define the likelihood function for the nonlinear regression model:

$$L(\beta, \sigma^2) = \frac{1}{(2\pi\sigma^2)^{\left(\frac{n}{2}\right)}} \exp\left\{-\frac{\sum_{i=1}^{n}\left[y_i - f(\beta, x)\right]^2}{2\sigma^2}\right\}$$

Step III: Maximize the likelihood function by minimizing the following:

$$S(\beta) = \sum_{i=1}^{n}\left[y_i - f(\beta, x)\right]^2$$

Step IV: Differentiate $S(\beta)$:

$$\frac{\partial S(\beta)}{\partial \beta} = -2\sum_{i=1}^{n}\left[y_i - f(\beta, x)\right]\frac{\partial f(\beta)}{\partial \beta}$$

Step V: Solve for the model parameters:
The parameters are then estimated by setting the partial derivatives equal to zero. This is determined via maximum likelihood estimation techniques.

Fig. 2.3 illustrates a nonlinear regression model for the different explanatory variables: order size, volatility, price, and percentage of volume (POV) rates. The graph shows how the dependent variable market impact cost varies with different values for the explanatory variables. This type of nonlinear model is extremely important for electronic trading and has become the underlying foundation for trading algorithms, high-frequency trading strategies, and portfolio construction.

2.7 CONCLUSIONS

In this chapter we provided an overview of regression models that can be used as the basis for sports prediction. These models are used to predict the winning team, estimate the winning score (home team points minus away team points), and to also calculate the probability of winning.

We provided an overview of three different types of regression models, including linear regression, log-linear regression, and nonlinear regression models.

These techniques provide analysts with methods to (1) determine the relationship between a dependent variable Y and a set of explanatory factors, (2) estimate parameter values, (3) calculation output variable Y, and (4) performance sensitivity and what-if analysis.

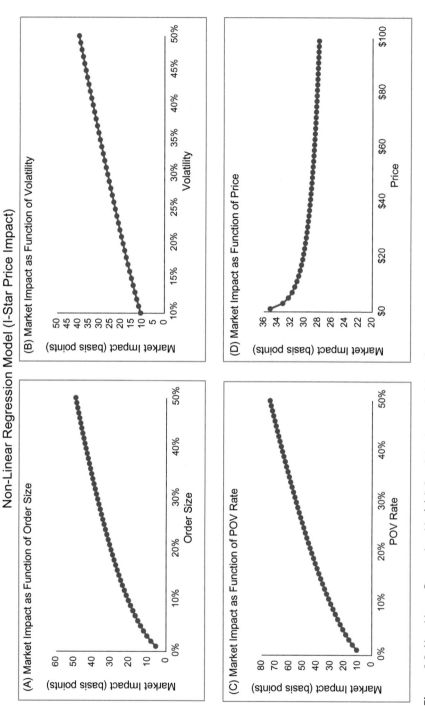

Figure 2.3 Non-Linear Regression Model (I-Star Price Impact). (A) Market Impact as Function of Order Size, (B) Market Impact as Function of Volatility, (C) Market Impact as Function of POV Rate, (D) Market Impact as Function of Price.

These techniques allow analysts to determine the likely winner of a game, the expected home winning margin, and the probability that the home team will win the game.

REFERENCES

Fox, J. (2002). *Nonlinear Regression and Nonlinear Least Squares: Appendix to an R and S-Plus companion to applied regression.* <http://cran.r-project.org/doc/contrib/Fox-Companion/appendix-nonlinear-regression.pdf>.

Greene, W. (2000). *Econometric Analysis* (4th ed.). Upper Saddle River, NJ: Prentice-Hall, Inc.

Gujarati, D. (1988). *Basic Economics* (2nd ed.). New York, NY: McGraw-Hill.

Kennedy, P. (1998). *A Guide to Econometrics* (4th ed.). Cambridge, MA: The MIT Press.

Kissell, R. (2013). *The Science of Algorithmic Trading and Portfolio Management.* Amsterdam: Elsevier/Academic Press.

Kissell, R., Glantz, M., & Malamut, R. (2004). A practical framework for estimating transaction costs and developing optimal trading strategies to achieve best execution. *Finance Research Letters, 1*, 35−46.

Kissell, R., & Malamut, R. (1999). *Optimal Trading Models.* Working Paper.

Kissell, R., & Malamut, R. (2006). Algorithmic decision making framework. *Journal of Trading, 1*(1), 12−21.

Zhi, J., Melia, A. T., Guericiolini, R., et al. (1994). Retrospective population-based analysis of the dose-response (fecal fat excretion) relationship of Orlistat in normal and obese volunteers. *Clinical Pharmacology and Therapeutics, 56*, 82−85.

CHAPTER 3

Probability Models

3.1 INTRODUCTION

In this chapter we provide an overview of mathematical probability models that can be used to rank teams, predict the winner of a game or match, and estimate the winning margin and the total number of points likely to be scored.

Different prediction methods require different input and output data. One of the most important tasks of data analysis is determining the most important set of input data that will provide an accurate estimate of the outcome variable you are trying to predict, such as winning team or score. Additionally, it is important to use only an appropriate number of input data to avoid overcomplication of the model. For example, many calculated data sets, especially in sports, are highly correlated with other data sets already being used. In these cases, incorporation of additional "correlated" data items will not help improve the prediction accuracy of the model and it may in fact lead to incorrect inferences and erroneous conclusions.

In this chapter we provide the proper techniques to develop, evaluate, and utilize probability prediction models.

3.2 DATA STATISTICS

Sports modeling data needs can be broken into two sets of data: input data (explanatory data) and output data (predicted events). In all of our modeling cases, we need to (1) determine the proper statistical relationship between explanatory factors and the output variable we are trying to predict, and (2) determine the parameter values that describe the underlying relationship between the input data and output variable. This will allow us to predict the outcome with the highest degree of statistical accuracy possible.

The *input data* are commonly known as the *x*-variables, right-hand side (RHS) variables, input factors, explanatory factors, independent variables, and/or predictor variables. In all cases, these terms mean exactly the same thing and consist of the data that will be used to predict the outcome.

Optimal Sports Math, Statistics, and Fantasy.
DOI: http://dx.doi.org/10.1016/B978-0-12-805163-4.00003-7

In probability models, the explanatory factors can be comprised of team data statistics or based on derived statistical data such as a team strength rating, and/or an offensive rating score and a defensive rating score. In many situations, we find that using derived data statistics consisting of team strength ratings provides superior results over actual calculated data items such as yards per rush and/or yards per pass in football.

The *output data* used in sports models represent the outcome event that we are trying to predict. These output data are commonly referred to as *y*-variables, left-hand side (LHS) data, dependent variable, outcome, predictor, and/or expected value.

In sports analysis, the output data that we are primarily trying to predict can be categorized into four sets of data: (1) win/lose, (2) winning margin or spread, (3) total points scored, and (4) player performance statistics. Results from these models will allow us to calculate additional items such as team rankings, probability of winning, probability of betting against a sports line, etc. Many of these topics will be further developed in later chapters.

A description of our output data items is as follows:

Win/Lose: A binary data value where 1 designates the home team won and 0 designates that the home team lost. In the case of a tie, analysts can include two input records for the game where the home team is denoted as the winner $+1$ in the first record and the home team denoted as the loser in the second record.

In our approach throughout this book, we will specify whether a team has won or lost from the perspective of the home team. If the home team won the game then the output variable is $+1$ and if the home team lost the game then the output variable is 0.

Spread/Margin: The spread, also known as the margin or winning margin, denotes the difference in score between the home team and away team. It is important to note here that we compute the home team winning spread based on the difference between the home team scoring 24 points and the visiting team scoring 21. That is, the home team predicted winning margin is $24 - 21 = +3$, thus, indicating the home team is expected to win 3 points. If the home team scored 21 points and the visiting team scored 24 points, then the winning margin for the home team is $21 - 24 = -3$, thus indicating that home team lost by 3 points. The winning margin can take on any value.

Total Points: This refers to the total points scored in a game or contest and is the sum of the points scored by the home team and by the visiting team. For example, if the home team scored 24 points and the visiting team scored 21 points then the total points scored for the game is $24 + 21 = 45$. The total points scored outcome variable cannot be negative. That is, it can be any number $>= 0$. Our models can also be analyzed based on the points scored by the home team only or based on the points scored by the away team. In these scenarios, we can predict the winner of the game based on the difference between expected points scored by the home team and expected points scored by the away team.

Player Performance: Player performance statistics are data results that describe the player's performance during a game or over a season. This can include ratio of hits to at-bats for a batter, number of earned runs allowed for a pitcher, number of yards rushed or points scored for a running back, number of passing yards or passing completion rate for a quarterback, points scored or shooting percentage for a point guard, etc. It is most important that the statistics used to describe player performance are those metrics that are most predictive of a team's winning probability. An in the case of fantasy sports, it is essential that the statistics being used to predict the expected points from a player are consistent with the scoring rules of the fantasy sports contest. Different fantasy sports competitions, even for the same sport and run by the same organization, may have different scoring systems.

Derived Data: Output data is data that is derived from different statistics. It can be an average of two different estimates or it can be the results of a different model or methodology. For example, in the logit regression below, we fit a model using probability data that is derived from the logistic function using different sampling techniques.

3.3 FORECASTING MODELS

The sections below provide examples of different sports forecasting models that can be used to predict winning percentage, winning margin, and points.

3.4 PROBABILITY MODELS

A probability model is a model where the outcome value can take on any value between zero and one. In most of these cases, we are estimating the probability that one team will beat another team. But the models can also incorporate a cumulative score based on the cumulative distribution function (CDF) of data statistics such as winning spread or points.

The trick behind incorporating probability models is as follows. First, represent the outcome variable as a value between zero and one. Second, incorporate a probability mapping function to ensure the prediction model will only result in outcome values between zero and one. These models are further discussed below.

For example, these models may be used to determine the probability that the home team A will beat the visiting team B in a game. This is written mathematically as:

$$P(A > B) = f(A, B) = p$$

In this notation, the expression $f(A, B)$ represents the model (formula) that will be used to estimate the probability p that team A will be victorious over team B in the event.

To utilize this modeling approach, we need to define three different items: model formula, input data, and output data.

Model Formula: The probability model formula refers to the equation that will be used to calculate the expected outcome probability p. Our work will focus on two different probability models: power function and exponential function. These are:

- Power function: $f(x) = \dfrac{x_1}{x_1 + x_2}$

- Logistic function: $f(x) = \dfrac{1}{(1 + e^{-(x_1 - x_2)})}$

Input Data: The input data used in our examples will consist of a derived team strength metric. This team strength metric is also known as the team rating. The terms *strength* and *rating* will be used interchangeably throughout the book. Analysts can apply techniques in this chapter to derive an overall team rating, and also an offensive and defensive team rating score.

Output Data: The output data used in probability models requires the output value y to be defined on the interval between 0 and 1. We provide examples below based on a binary win/loss metric, and we also include examples showing how an analyst can transform the margin or points scored to values between 0 and 1 using various distribution functions.

Determining the proper output data to use in a probability model is often a very difficult task. This is because we do not know with certainty the exact likelihood that team A will beat team B and it is very difficult to ascertain exact probabilities from a limited number of game observations. For example, suppose that the exact probability that team A will beat team B is 80%. This implies that we should observe game results between A and B where team A indeed beats team B in 8 out of 10 games. But now, if we only have a single observation of results between A and B we will observe that either A beat B (the 80% chance) or that B beat A (the 20% chance).

Our data set of game results does not provide enough information to determine that the true probability is 80%. Additionally, if the probability that C beat D is 50% and we have three game observations we will find that one of the teams, say C, won at least two games and possibly three while the other team, say D, lost two games and possibility three.

By grouping observations, we will begin to calculate the probability that C wins over D to be $2/3 = 67\%$ or $3/3 = 100\%$. In both cases, the results are far from the exact probability. The key point here is that if we are trying to compute probability levels from game results we need enough observations between the teams (i.e., a statistically sufficient data set) to have accurate probability levels. Unfortunately, in most sports we do not have enough observations across teams and additionally, we do not have observations across all combinations of teams.

To determine "true" probability levels from head-to-head matchups we need a much larger number of observations and game results. But since we typically only have a small sample of games we need to employ statistical analysis to compute these probabilities.

To resolve this limited data issue, we can solve our models using maximum likelihood estimation (MLE) techniques based solely on if the team won or lost a game. That is, the outcome variable will be denoted as 1 if the team won and 0 if the team lost.

Another important modeling item with respect to probability models is that we need to evaluate the error term of our estimation model. In many cases, our conclusions and predictions are based on having an error term that is normally distributed. If the error term (e.g., model noise) follows a different distribution, then we need to consider these probabilities in our predictions. In Chapter 4, Advanced Math and Statistics, we discuss some of the more important probability distributions for sports models.

Example 3.1

Consider a six team sporting event where each team plays a series of games against the other teams. The outcome of these games is shown in the digraph in Fig. 3.1, where a line indicates a game between the two teams and an arrow designates the winner of

Games Results

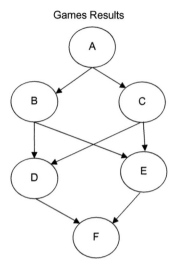

Figure 3.1 Games Results.

the game. For example, there is a line between team A and team B with the arrow pointing to team B. This indicates that team A beat team B.

Based on the digraph and set of results, an obvious ranking of these teams is A #1, B&C (tied for 2nd & 3rd), D&E (tied for 4th & 5th), and F (last).

We can also use the MLE process to estimate the set of parameter values that will maximize the likelihood of observing all outcomes and thus provide a ranking of teams. This approach will prove to be more beneficial especially when there is a much larger number of games and when there is no clear or obvious ranking of teams.

These steps are as follows:

Step 1. Define power function formula:

$$F(X > Y) = \frac{X}{X + Y}$$

Step 2. Set up a likelihood function based on observed outcomes.

$$L = P(A > B) \cdot P(A > C) \cdot P(B > D) \cdot P(B > E) \cdot P(C > D) \cdot P(C > E) \cdot P(D > F) \cdot P(E > F)$$

Step 3. Write the likelihood function in terms of the power function formula:

$$L = \frac{A}{A + B} \cdot \frac{A}{A + C} \cdot \frac{B}{B + D} \cdot \frac{B}{B + E} \cdot \frac{C}{C + D} \cdot \frac{C}{C + E} \cdot \frac{D}{D + F} \cdot \frac{E}{E + F}$$

Unfortunately, when we set out to solve a multiplicative expression where each term is less than one the product of all these values falls to zero exceptionally quickly, thus making it extremely difficult to solve the equation above. For example, if we multiply 0.5 ten times (i.e., there are ten games), 0.50^10 − 0.000977, the result is extremely small. This value L also becomes exponentially smaller as the number of games increases.

A solution around this data issue is to transform the above equation into one using logs and thus turn a difficult-to-solve multiplicative expression into an easier-to-solve additive expression. Additionally, it is important to note that maximizing the log transformation of this function L will yield the same results as if we maximize the actual function L.

Step 4. Transform the likelihood function into a log-likelihood function:

$$\log L = \ln\left(\frac{A}{A+B}\right) + \ln\left(\frac{A}{A+C}\right) + \ln\left(\frac{B}{B+D}\right) + \ln\left(\frac{B}{B+E}\right) + \ln\left(\frac{C}{C+E}\right)$$

$$+ \ln\left(\frac{C}{C+F}\right) + \ln\left(\frac{D}{D+F}\right) + \ln\left(\frac{E}{E+F}\right)$$

Now, our problem has turned into one where we simply need to determine the parameter values A, B, C, D, E, and F that maximize our log-likelihood function $\log L$.

This maximization optimization problem requires bounds to be placed on the maximum and minimum parameter values. Otherwise, the solution will make the parameter value for the team that only wins approach infinity and the team that only loses approach zero.

Step 5. Set up the maximization optimization with lower and upper bounds.

$$\log L = \ln\left(\frac{A}{A+B}\right) + \ln\left(\frac{A}{A+C}\right) + \ln\left(\frac{B}{B+D}\right) + \ln\left(\frac{B}{B+E}\right) + \ln\left(\frac{C}{C+E}\right)$$

$$+ \ln\left(\frac{C}{C+F}\right) + \ln\left(\frac{D}{D+F}\right) + \ln\left(\frac{E}{E+F}\right)$$

s.t.,

$$0 < A \leq 5$$
$$0 < B \leq 5$$
$$0 < C \leq 5$$
$$0 < D \leq 5$$
$$0 < E \leq 5$$
$$0 < F \leq 5$$

It is important to note that in the above formulation, the parameter value cannot take on a value of zero because we cannot have a value of zero in the denominator (in the event both team parameters are zero) and the log of zero in undefined (in the event that the winning team has a parameter of zero and appears in the numerator). Thus, the values of our parameters need to be strictly greater than the lower bound zero but less than or equal to the upper bound 5.

Step 6. Solve for the parameters

The parameters that maximize $\log L$ can be determined using Excel's Solver function or another statistical software application such as MATLAB.

Table 3.1A Optimization Formulation

	Game Results			Initialization Phase				Final Solution Phase			
Obs	Win Team	Lose Team		Win Parameter	Lose Parameter	Initial Fx	Initial Ln Fx	Win Parameter	Lose Parameter	Final Fx	Final Ln Fx
1	A	B		2.500	2.500	0.500	− 0.693	5.000	0.373	0.931	− 0.072
2	A	C		2.500	2.500	0.500	− 0.693	5.000	0.373	0.931	− 0.072
3	B	D		2.500	2.500	0.500	− 0.693	0.373	0.013	0.965	− 0.035
4	B	E		2.500	2.500	0.500	− 0.693	0.373	0.013	0.965	− 0.035
5	C	D		2.500	2.500	0.500	− 0.693	0.373	0.013	0.965	− 0.035
6	C	E		2.500	2.500	0.500	− 0.693	0.373	0.013	0.965	− 0.035
7	D	F		2.500	2.500	0.500	− 0.693	0.013	0.001	0.931	− 0.072
8	E	F		2.500	2.500	0.500	− 0.693	0.013	0.001	0.931	− 0.072
						Initial Log $L =$	− 5.545			Final Log $L =$	− 0.429

Table 3.1B Optimization Parameter Values

	Parameter Values				Final Rankings	
Team	Initial Parameter	Final Parameter	LB	UB	Team Ranking	Parameter Value
A	2.500	5.000	0.001	5.000	A	5.000
B	2.500	0.373	0.001	5.000	C	0.373
C	2.500	0.373	0.001	5.000	B	0.373
D	2.500	0.013	0.001	5.000	E	0.013
E	2.500	0.013	0.001	5.000	D	0.013
F	2.500	0.001	0.001	5.000	F	0.001

The model setup is shown in Table 3.1A. The table shows both the initial values used at the start of the optimization routine and the final optimized solution parameters. The parameter constraints are shown in Table 3.1B.

In Table 3.1A we separate the parameter and function values into the initialization phase and the final solution phase for illustrative purposes only. In the initialization phase we set the parameter values to a starting value of 2.5 for all teams as a starting value and to show how the initial probabilities and objective function value is determined. Thus the probability p that the winning teams wins a game is:

$$p = \frac{2.5}{2.5 + 2.5} = .50$$

This is noted in the "Initial Fx" column = 0.50 and the log transformation of this value is Ln Fx = −0.693. The goal of the optimization process is to maximize the sum of the Ln Fx across all games. This sum based on the initial parameter value of 2.5 across all teams is −5.545.

The final solution section shows the optimized team parameters after the MLE optimization process. Notice now that the final Log L value is −0.429 and increased from −5.545 to −0.429. The final team parameter values and team rankings after performing the optimization are shown in Table 3.1B. These are A = 5.0, B = 0.373 & C = 0.373, D = 0.013, & E = 0.013, and F = 0.001. Notice that these parameter values and corresponding team rankings are consistent with the intuitive observed rankings determined from Fig. 3.1.

A benefit of having optimized team parameter values is that it allows us to make comparisons and predictions across any two teams even if these teams have not played each other and even if these teams do not have any common opponents. The optimization process is robust enough to evaluate every connect across teams, thus providing final insight to compare two teams based on head-to-head games, based on performance against common opponents, and even based on common opponents of common opponents.

Table 3.1C provides the probability that team X in the left-hand column will beat team Y shown on the rows of the table. The probabilities are computed based on the

Table 3.1C Probability Estimates						
Probability Team X (LHS Rows) will Beat Team Y (Top Columns)						
X/Y	A	B	C	D	E	F
A	—	93%	93%	99%	99%	99%
B	7%	—	50%	97%	97%	99%
C	7%	50%	—	97%	97%	99%
D	0%	3%	3%	—	50%	93%
E	0%	3%	3%	50%	—	93%
F	0%	0%	0%	7%	7%	—

optimized parameter values and allow us to make a "comparison" across teams that did play one another (such as team A and team B) and also for teams that did not play one another (such as team A & team F). These probabilities are computed based on the optimized parameter values as follows:

$$Prob(X > Y) = \frac{V(x)}{V(x) + V(y)}$$

where $V(x)$ and $V(y)$ represent the optimized parameter values for team X and team Y respectively. Thus, the probability that team A will beat team D is computed as follows:

$$Prob(A > D) = \frac{5.0}{5.0 + 0.013} = .997 = 99.7\%$$

Example 3.2

Now let us consider the same example but with an additional team G. In this scenario, team G has beaten teams D, E, and F. The digraph showing these games and results is shown in Fig. 3.2. Notice in the diagraph that all we know about team G is that they should be ranked higher than teams D, E, and F. But how much higher should they be ranked? That is, from these results alone, can we determine that team G is as good as team A and should be ranked tied with team A? Or should they be ranked at the same level as teams B & C? Or possibly, ranked somewhere between A and B&C or between B & C and just above D & E?

This is the very issue that many ranking committees face each year when ranking teams for a top 25 poll or trying to rank teams for inclusions in a postseason tournament. Very often we hear comments that a team that only plays weaker opponents and is undefeated should not rank at the top. But is this correct? In this example, there is no data for us to conclude that team G should be ranked tied for #1 with team A, be ranked equal with B&C, or in any position that is above D&E. An objective model that

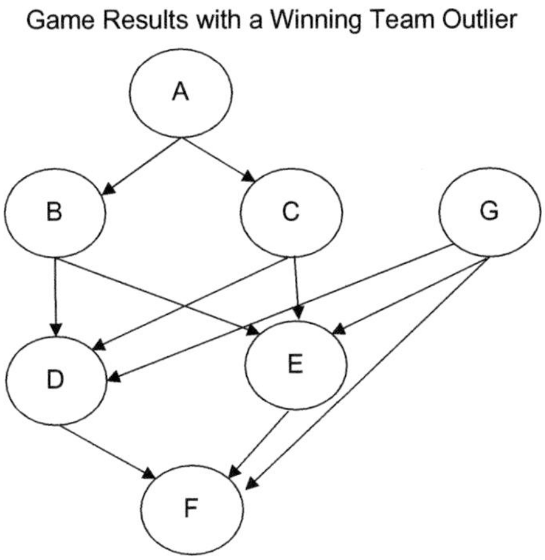

Game Results with a Winning Team Outlier

Figure 3.2 Game Results with a Winning Team Outlier.

seeks to maximize the probability of game results based only on whether a team has won or lost a game would have to rank all undefeated teams tied for #1 and all teams with only losses as tied for last. This is the very issue we have had with many of the sports ranking models in use for the previous BCS tournament and any top 25 poll that only allows a team to be ranked if it won or lost the game and cannot consider the final score in the rankings.

In the past when we inquired about some of these models in use that do not rank undefeated teams as #1, the operators of these models state that they may be doing some postoptimization adjustment to correct for these issues. (But this seems to violate the objective nature of the sports model—at least to us.)

A model should only be used by a committee for postseason selection or for top 25 rankings if and only if the model is transparent and can be duplicated and verified, it has sound methodology that is agreed to as the appropriate methodology to use for the rankings or selection, and the model remains completely objective and does not perform any postoptimization process to correct for issues. For example, what would make someone believe a model result should be corrected if they do not have some subjective bias to begin with regarding the ranking?

A probability-maximizing model should always seek to maximize the likelihood of observing all sets of outcomes, and should not make any adjustments. Criticism regarding this approach is that we may result in teams being ranked higher than they actually should since we are pushing all undefeated teams to the top. While is this true, selection committees should rely on additional methodology as well to help rank these teams.

The result of our power function probability-maximizing optimization results in team G being ranked tied with team A. These power ratings are #1 A&G (5.0), #3 B&C (0.368), #5 D&E (0.13), #7 F (0.001).

Table 3.2A Optimization Formulation

	Game Results		Initialization Phase				Final Solution Phase			
Obs	Win Team	Lose Team	Win Parameter	Lose Parameter	Initial Fx	Initial Ln Fx	Win Parameter	Lose Parameter	Final Fx	Final Ln Fx
1	A	B	2.500	2.500	0.500	−0.693	5.000	0.368	0.931	−0.071
2	A	C	2.500	2.500	0.500	−0.693	5.000	0.368	0.931	−0.071
3	B	D	2.500	2.500	0.500	−0.693	0.368	0.013	0.966	−0.035
4	B	E	2.500	2.500	0.500	−0.693	0.368	0.013	0.966	−0.035
5	C	D	2.500	2.500	0.500	−0.693	0.368	0.013	0.966	−0.035
6	C	E	2.500	2.500	0.500	−0.693	0.368	0.013	0.966	−0.035
7	D	F	2.500	2.500	0.500	−0.693	0.013	0.001	0.929	−0.074
8	E	F	2.500	2.500	0.500	−0.693	0.013	0.001	0.929	−0.074
9	G	D	2.500	2.500	0.500	−0.693	5.000	0.013	0.997	−0.003
10	G	E	2.500	2.500	0.500	−0.693	5.000	0.013	0.997	−0.003
11	G	F	2.500	2.500	0.500	−0.693	5.000	0.001	1.000	0.000
					Initial Log L =	−7.62462			Final Log L =	−0.435

Table 3.2B Optimization Parameter Values

	Parameter Values				Final Rankings		
Team	Initial Parameter	Final Parameter	LB	UB	Team	Team Ranking	Parameter Value
A	2.500	5.000	0.001	5.000	A	1	5.000
B	2.500	0.368	0.001	5.000	B	4	0.368
C	2.500	0.368	0.001	5.000	C	3	0.368
D	2.500	0.013	0.001	5.000	D	6	0.013
E	2.500	0.013	0.001	5.000	E	5	0.013
F	2.500	0.001	0.001	5.000	F	7	0.001
G	2.500	5.000	0.001	5.000	G	1	5.000

Table 3.2A shows the initial and final optimization parameters for this example. Table 3.2B shows the optimization lower and upper bounds, and the final rankings for each team.

Example 3.3

Now let us consider the same scenario as in Example 3.1 but with an additional team H that has lost to teams A, B, and C. The digraph showing these games and results is shown in Tables 3.3A and 3.3B. Notice in the diagraph that all we know about team G is that they should be ranked lower than A, B, and C. But how much lower? Are they at the bottom of the rankings with team F or should they be at the same level as teams E and F? Similar to the issues in Example 3.2 with a team that has only beaten weaker teams, we now face an issue where a team has lost to all highly ranked and stronger teams (Fig. 3.3).

Unfortunately, there is not enough data in this scenario for us to determine exactly where team H should rank in comparison to the other team. Our probability-maximizing model, hence, finds that team H will be rated at the bottom and on par with team F.

These final rankings from our power function optimization are #1 A (5.0), #2 B&C (0.368), #4 D&E (0.13), #6 F&H (0.001).

Table 3.3A shows the initial and final optimization parameters for this example. Table 3.3B shows the optimization lower and upper bounds, and the final rankings for each team.

Example 3.4

Another example that must be discussed from the perspective of the probability model is that of circular logic. Many times we find a situation where three teams have played one other with A beating B, B beating C, and C beating A (see Fig. 3.4). But the rankings of these teams is A#1, B#2, and C#3 because the ranking procedure being used is subjective and may be based on history or perception, which should not factor into the present rankings.

Table 3.3A Optimization Formulation

	Game Results			Initialization Phase				Final Solution Phase			
Obs	Win Team	Lose Team	Win Parameter	Lose Parameter	Initial Fx	Initial Ln Fx	Win Parameter	Lose Parameter	Final Fx	Final Ln Fx
1	A	B	2.500	2.500	0.500	−0.693	5.000	0.383	0.929	−0.074
2	A	C	2.500	2.500	0.500	−0.693	5.000	0.383	0.929	−0.074
3	B	D	2.500	2.500	0.500	−0.693	0.383	0.014	0.966	−0.035
4	B	E	2.500	2.500	0.500	−0.693	0.383	0.014	0.966	−0.035
5	C	D	2.500	2.500	0.500	−0.693	0.383	0.014	0.966	−0.035
6	C	E	2.500	2.500	0.500	−0.693	0.383	0.014	0.966	−0.035
7	D	F	2.500	2.500	0.500	−0.693	0.014	0.001	0.931	−0.071
8	E	F	2.500	2.500	0.500	−0.693	0.014	0.001	0.931	−0.071
9	A	H	2.500	2.500	0.500	−0.693	5.000	0.001	1.000	0.000
10	B	H	2.500	2.500	0.500	−0.693	0.383	0.001	0.997	−0.003
11	C	H	2.500	2.500	0.500	−0.693	0.383	0.001	0.997	−0.003
					Initial Log L =	−7.625			Final Log L =	−0.435

Table 3.3B Optimization Parameter Values

	Parameter Values				Final Rankings		
Team	Initial Parameter	Final Parameter	LB	UB	Team	Team Ranking	Parameter Value
A	2.500	5.000	0.001	5.000	A	1	5.000
B	2.500	0.383	0.001	5.000	B	4	0.383
C	2.500	0.383	0.001	5.000	C	3	0.383
D	2.500	0.014	0.001	5.000	D	6	0.014
E	2.500	0.014	0.001	5.000	E	5	0.014
F	2.500	0.001	0.001	5.000	F	7	0.001
G	2.500	5.000	0.001	5.000	G	1	5.000
H	2.500	0.001	0.001	5.000	H	8	0.001

Game Results with a Losing Team Outlier

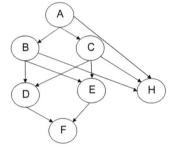

Figure 3.3 Game Results with a Losing Team Outlier.

Game Results with Circular Outlier

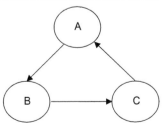

Figure 3.4 Game Results with Circular Outlier.

Table 3.4A Optimization Formulation

	Game Results			Initialization Phase				Final Solution Phase			
Obs	Win Team	Lose Team		Win Parameter	Lose Parameter	Initial (Fx)	Initial Ln (Fx)	Win Parameter	Lose Parameter	Final (Fx)	Final Ln (Fx)
1	A	B		2.500	2.500	0.500	−0.693	2.500	2.500	0.500	−0.693
2	B	C		2.500	2.500	0.500	−0.693	2.500	2.500	0.500	−0.693
3	C	A		2.500	2.500	0.500	−0.693	2.500	2.500	0.500	−0.693
						Initial Log $L =$	−2.079			Final Log $L =$	−2.079

Table 3.4B Optimization Parameter Values

Parameter Values				Final Rankings		
Team	Initial Parameter	Final Parameter	LB	UB	Team	Team Ranking	Parameter Value
A	2.500	2.500	0.001	5.000	A	2.500	1.000
B	2.500	2.500	0.001	5.000	B	2.500	1.000
C	2.500	2.500	0.001	5.000	C	2.500	1.000

For example, if in a basketball contest Duke beats Michigan, Michigan beats Stony Brook, and Stony Brook beats Duke, a nonprobability ranking of the team may rank Duke #1, Michigan #2, followed by Stony Brook #3 based on perception or belief of the strength of each team. But there is not any evidence in the empirical win/loss data that would suggest that there is a difference between the strength of these teams. Thus, all teams should be ranked equal to one another.

It is safe to state that our probability-rating optimization approach results in all three teams being rated exactly the same in this situation. These ratings are A = 5.0, B = 5.0, and C = 5.0.

Table 3.4A shows the initial and final optimization parameters for this example. Table 3.4B shows the optimization lower and upper bounds, and the final rankings for each team.

Example 3.5 Logistic Model Optimization

We next demonstrate the use of a logistic probability optimization model to rank the ten (10) teams from the example used in Chapter 2, Regression Models. This example has 10 teams with each team playing every other team both home and away. Thus, each team plays 18 games (9 home games and 9 away games) and there are 90 games in total. These results are shown in Table 3.5A.

The only difference between the power function and exponential function technique is the mapping of parameter value to the probability space (values between 0 and 1). In most cases, these resulting rankings between models will be highly correlated but the probability levels corresponding to winning probability level will at times be different due to the function analyzed. Thus, usage of both models provides a secondary level of insight and hence further confidence or lack of confidence surrounding our estimates. For example, if our two approaches estimate the probability that X beat Y to be 92% and 95% we are certain that X has a very high likelihood of winning the game (i.e., >90%). But if one approach states a 55% likelihood of winning and the other approach a 45% then we are not as confident that X will beat Y.

Table 3.5A Game Results

Game Number	Home Team	Away Team	Winning Team	Losing Team	Home Spread	Actual Probability	Est. Logistic Probability	Spread Z-Score	Spread CDF
1	1	2	1	2	6	0.947	0.996	0.449	0.673
2	1	3	1	3	1	0.340	0.442	−0.349	0.363
3	1	4	1	4	7	0.860	0.988	0.608	0.728
4	1	5	5	1	−1	0.512	0.659	−0.669	0.252
5	1	6	1	6	1	0.639	0.754	−0.349	0.363
6	1	7	7	1	−3	0.267	0.338	−0.988	0.162
7	1	8	1	8	1	0.120	0.107	−0.349	0.363
8	1	9	9	1	−1	0.510	0.659	−0.669	0.252
9	1	10	10	1	−7	0.068	0.061	−1.626	0.052
10	2	1	1	2	−1	0.100	0.098	−0.669	0.252
11	2	3	3	2	−6	0.028	0.017	−1.467	0.071
12	2	4	2	4	3	0.463	0.643	−0.030	0.488
13	2	5	5	2	−1	0.049	0.041	−0.669	0.252
14	2	6	6	2	−3	0.080	0.063	−0.988	0.162
15	2	7	7	2	−1	0.019	0.011	−0.669	0.252
16	2	8	8	2	−6	0.009	0.003	−1.467	0.071
17	2	9	9	2	−3	0.052	0.041	−0.988	0.162
18	2	10	10	2	−4	0.005	0.001	−1.147	0.126
19	3	1	3	1	10	0.833	0.968	1.087	0.862
20	3	2	3	2	7	0.921	0.999	0.608	0.728
21	3	4	3	4	10	0.925	0.998	1.087	0.862
22	3	5	3	5	7	0.757	0.923	0.608	0.728
23	3	6	3	6	7	0.835	0.950	0.608	0.728
24	3	7	3	7	6	0.565	0.760	0.449	0.673
25	3	8	8	3	−1	0.351	0.427	−0.669	0.252
26	3	9	3	9	9	0.738	0.923	0.927	0.823
27	3	10	10	3	−1	0.221	0.287	−0.669	0.252
28	4	1	1	4	−2	0.212	0.228	−0.828	0.204
29	4	2	4	2	7	0.800	0.930	0.608	0.728
30	4	3	3	4	−2	0.063	0.046	−0.828	0.204
31	4	5	5	4	−2	0.113	0.104	−0.828	0.204
32	4	6	4	6	1	0.175	0.156	−0.349	0.363
33	4	7	7	4	−7	0.045	0.030	−1.626	0.052
34	4	8	8	4	−13	0.021	0.007	−2.584	0.005
35	4	9	9	4	−4	0.115	0.104	−1.147	0.126
36	4	10	10	4	−11	0.011	0.004	−2.265	0.012
37	5	1	5	1	9	0.750	0.926	0.927	0.823
38	5	2	5	2	9	0.967	0.998	0.927	0.823
39	5	3	3	5	−4	0.544	0.668	−1.147	0.126
40	5	4	5	4	10	0.937	0.995	1.087	0.862
41	5	6	5	6	6	0.775	0.886	0.449	0.673
42	5	7	5	7	3	0.429	0.564	−0.030	0.488
43	5	8	8	5	−4	0.215	0.234	−1.147	0.126
44	5	9	9	5	−6	0.664	0.831	−1.467	0.071
45	5	10	10	5	−5	0.132	0.141	−1.307	0.096

(Continued)

Table 3.5A (Continued)

Game Number	Home Team	Away Team	Winning Team	Losing Team	Home Spread	Actual Probability	Est. Logistic Probability	Spread Z-Score	Spread CDF
46	6	1	6	1	5	0.647	0.887	0.289	0.614
47	6	2	6	2	5	0.871	0.997	0.289	0.614
48	6	3	6	3	5	0.389	0.559	0.289	0.614
49	6	4	6	4	10	0.836	0.992	1.087	0.862
50	6	5	6	5	4	0.519	0.756	0.129	0.552
51	6	7	6	7	1	0.323	0.449	−0.349	0.363
52	6	8	8	6	− 3	0.142	0.161	−0.988	0.162
53	6	9	6	9	9	0.586	0.756	0.927	0.823
54	6	10	10	6	− 3	0.084	0.094	−0.988	0.162
55	7	1	7	1	6	0.902	0.979	0.449	0.673
56	7	2	7	2	9	0.965	1.000	0.927	0.823
57	7	3	7	3	2	0.739	0.884	−0.190	0.425
58	7	4	7	4	12	0.955	0.999	1.406	0.920
59	7	5	7	5	11	0.830	0.949	1.247	0.894
60	7	6	7	6	10	0.822	0.967	1.087	0.862
61	7	8	7	8	1	0.444	0.536	−0.349	0.363
62	7	9	7	9	9	0.783	0.949	0.927	0.823
63	7	10	10	7	− 1	0.304	0.384	−0.669	0.252
64	8	1	8	1	10	0.868	0.995	1.087	0.862
65	8	2	8	2	10	0.916	1.000	1.087	0.862
66	8	3	8	3	4	0.822	0.970	0.129	0.552
67	8	4	8	4	15	0.916	1.000	1.885	0.970
68	8	5	8	5	17	0.841	0.987	2.204	0.986
69	8	6	8	6	10	0.914	0.992	1.087	0.862
70	8	7	8	7	14	0.801	0.954	1.726	0.958
71	8	9	8	9	10	0.836	0.987	1.087	0.862
72	8	10	8	10	6	0.522	0.725	0.449	0.673
73	9	1	9	1	6	0.785	0.926	0.449	0.673
74	9	2	9	2	5	0.900	0.998	0.289	0.614
75	9	3	3	9	− 2	0.538	0.668	−0.828	0.204
76	9	4	9	4	10	0.894	0.995	1.087	0.862
77	9	5	5	9	− 4	0.624	0.831	−1.147	0.126
78	9	6	9	6	1	0.773	0.886	−0.349	0.363
79	9	7	7	9	− 1	0.423	0.564	−0.669	0.252
80	9	8	8	9	− 5	0.205	0.234	−1.307	0.096
81	9	10	9	10	1	0.132	0.141	−0.349	0.363
82	10	1	10	1	10	0.954	0.997	1.087	0.862
83	10	2	10	2	13	0.991	1.000	1.566	0.941
84	10	3	10	3	8	0.873	0.984	0.768	0.779
85	10	4	10	4	11	0.905	1.000	1.247	0.894
86	10	5	10	5	11	0.892	0.993	1.247	0.894
87	10	6	10	6	4	0.955	0.996	0.129	0.552
88	10	7	10	7	2	0.815	0.975	−0.190	0.425
89	10	8	10	8	3	0.795	0.901	−0.030	0.488
90	10	9	10	9	5	0.943	0.993	0.289	0.614

The logistic function described above that calculates the probability that team X will beat team Y is defined as:

$$P(X > Y) = \frac{1}{1 + e^{-(x-y)}}$$

We expand on this notation to include a home-field advantage term and the probability that the home team will beat the away team is calculated as:

$$P(Home > Away) = \frac{1}{1 + e^{-(B0 + B\_Home - B\_Away)}}$$

This function ensures that the result will be between 0 and 1 as is required for a probability model, and also incorporates a home-field advantage parameter value to provide a potential benefit for the home team.

Since we can only observe if the home team or away team won the game, the formulation for each game result is:

$$f = \begin{cases} \dfrac{1}{1 + e^{-(B0 + B\_Home - B\_Away)}} & \textit{if Home Team Wins Game} \\[3ex] 1 - \dfrac{1}{1 + e^{-(B0 + B\_Home - B\_Away)}} & \textit{if Away Team Wins Game} \end{cases}$$

The formulation of the likelihood function L for the first five games in this season is:

$$L = \frac{1}{1 + e^{-(b0 + b1 - b2)}} \cdot \frac{1}{1 + e^{-(b0 + b1 - b3)}} \cdot \frac{1}{1 + e^{-(b0 + b1 - b4)}} \cdot \left(1 - \frac{1}{1 + e^{-(b0 + b1 - b5)}}\right)$$
$$\cdot \frac{1}{1 + e^{-(b0 + b1 - b6)}} \cdots$$

This model is solved via the same technique as above, that is, by taking the log transformation of these values and performing MLE on these results. The log-likelihood function for the first five games in the season is:

$$\text{Log } L = \ln\left(\frac{1}{1 + e^{-(b0 + b1 - b2)}}\right) + \ln\left(\frac{1}{1 + e^{-(b0 + b1 - b3)}}\right) + \ln\left(\frac{1}{1 + e^{-(b0 + b1 - b4)}}\right)$$
$$+ \ln\left(\left(1 - \frac{1}{1 + e^{-(b0 + b1 - b5)}}\right)\right) + \ln\left(\frac{1}{1 + e^{-(b0 + b1 - b6)}}\right) + \cdots$$

Notice that the home team won four of the first five games, losing only game 4. Therefore, only the fourth expression in the above log-likelihood example adjusts for the away team winning the game.

This model can be solved via Excel's Solver function or via a statistical package such as MATLAB and its optimization toolbox. But as the number of teams and number of games becomes larger and larger there is more of a need for a mathematical software such as MATLAB for the optimization due to the time to solve and mathematical precision needed.

Table 3.5B Logistic Probability Results and Team Ranking

Parameter Values				Team Ranking	
Team	Parameter	LB	UB	Team	Ranking
1	4.244	0.001	10.000	1	8
2	0.432	0.001	10.000	2	10
3	6.069	0.001	10.000	3	4
4	1.435	0.001	10.000	4	9
5	5.176	0.001	10.000	5	6
6	4.714	0.001	10.000	6	7
7	6.507	0.001	10.000	7	3
8	7.952	0.001	10.000	8	2
9	5.176	0.001	10.000	9	5
10	8.571	0.001	10.000	10	1
HFA	1.590	0.001	10.000		

Table 3.5C Probability that the Home Team will Beat the Away Team

Home/Away	1	2	3	4	5	6	7	8	9	10
1	—	99%	44%	99%	66%	75%	34%	11%	66%	6%
2	10%	—	2%	64%	4%	6%	1%	0%	4%	0%
3	97%	99%	—	99%	92%	95%	76%	43%	92%	29%
4	23%	93%	5%	—	10%	16%	3%	1%	10%	0%
5	93%	99%	67%	99%	—	89%	56%	23%	83%	14%
6	89%	99%	56%	99%	76%	—	45%	16%	76%	9%
7	98%	99%	88%	99%	95%	97%	—	54%	95%	38%
8	99%	99%	97%	99%	99%	99%	95%	—	99%	73%
9	93%	99%	67%	99%	83%	89%	56%	23%	—	14%
10	99%	99%	98%	99%	99%	99%	97%	90%	99%	—

We initialed the parameter value for this scenario at 5.0 with a lower bound of 0.001 and upper bound of 10.0 for each parameter. Notice that as we increase the number of teams and number of games we may need to increase the potential parameter set (with a larger upper bound) to achieve more precision. Proper selection of the upper bound in these models will come with experience as well as a thorough understanding of the mathematical techniques.

The rankings of teams and their corresponding rating values are in Table 3.5B.

Similar to the methods above, we can use our parameter values to compute the probability of winning between any two teams for both home and away games. These probability estimates across all 10 teams are described in the following sections (Table 3.5C).

3.5 LOGIT MODEL REGRESSION MODELS

A logit regression is a linearization of the logistic function described above. The logit model is an important and useful mathematical tool but does require the outcome variables to be between 0 and 1. In the examples above, our outcome variables were binary and could only take on the value of 0 or 1. For a logit regression we need these values to be between 0 and 1. When this is the case, the solution to the team ratings is straightforward and direct, and can be solved via ordinary least squares regression techniques.

For example, suppose that the probability of the home team beating the away team is known. In this case, the logit model will allow us to determine each team's strength "rating" and then apply these ratings across any pair of teams to predict the winner.

The logit regression probability model is solved via the following steps:

1. Specify the model functional form using the logistic equation:

$$P(x_1 > x_2) = \frac{1}{1 + e^{-(x_1 - x_2)}} = p$$

2. Calculate the wins-ratio by dividing the probably of a win by the probability of a loss. This is as follows:

$$Wins\ Ratio = \frac{P(x_1 > x_2)}{1 - P(x_1 > x_2)} = \frac{\frac{1}{1 + e^{-(x_1 - x_2)}}}{1 - \frac{1}{1 + e^{-(x_1 - x_2)}}} = \frac{p}{1 - p}$$

3. Simplify the expression:

$$e^{(x_1 - x_2)} = \frac{p}{1 - p}$$

4. Reduce using logs:

$$\ln\left(e^{(x_1 - x_2)}\right) = \ln\left(\frac{p}{1 - p}\right)$$

5. Which yields:

$$x_1 - x_2 = \ln\left(\frac{p}{1 - p}\right)$$

6. This transformation now allows us to rewrite our problem in terms of a linear regression model in the form:

$$y = Ab$$

where y is the log of the Wins–Ratio vector, A is the games matrix, and b is the vector of parameter values, which needs to be determined (i.e., the decision variable). The best way to illustrate the solution of this model is via example.

Example 3.6

We next demonstrate the solution of the logit regression model using games results data shown in Table 3.5A. This table also has the actual probability of the home team winning, which will serve as our y-variable. It is unfortunate, however, that in most cases we do not have the actual probability that one team will win the game, but the probability can be ascertained using sampling techniques that will be discussed in the next chapter and in an example below using the CDF of home team winning spread.

We compute the y-vector (LHS) and the A games matrix (RHS) using the data in Table 3.5A,. In general, if there are n-games and m-teams the y-vector will have dimension $(n + 1) \times 1$, and the A matrix will have dimension $(n + 1) \times (m + 1)$. So in our example with 10 teams and 90 total games, the y-vector will be 91×1 (i.e., 91 rows $\times$ 1 column) and the A matrix will be 91×11 (91 rows $\times$ 11 columns).

The A matrix will have a +1 in the column corresponding to the home team column and a −1 in the column corresponding to the away team column. The first column will be +1 to denote the constant term, which in this case represents the home-field advantage. All other entries will be zero.

The first five rows of the A matrix and the last row using data in Table 3.5A are shown in Table 3.6A.

Notice that the first five games consist of the home team $\times 1$ playing teams $\times 2$, $\times 3$, $\times 4$, $\times 5$, and $\times 6$. Therefore, the first column will have 1s to denote the constant term and home-field advantage, and the second column will also have 1s to denote home team $\times 1$ playing in each game. In the first row, we have −1 entered in the cell corresponding to the $\times 2$ team and the negative sign indicates $\times 2$ is the away team. In the second row we have −1 entered in the cell corresponding to the $\times 3$ team, again, the negative sign represents that the team is playing away (and so on).

Why is the last row of the input matrix all 1s? Readers familiar with matrix algebra will realize that the A matrix through construction will have a rank less than the number of columns, therefore making it very difficult to determine accurate parameter estimates.

In mathematical terminology, we say that the matrix has reduced rank. For example, if the matrix A does not have the last row of ones, it is a $n \times (m + 1)$ matrix and its rank is:

$$rank(A) = m < m + 1$$

Therefore, this results in all sorts of mathematical issues, which will most often result in inaccurate predictions and inaccurate rankings. This is one of the primary reasons why many

Table 3.6A Sample Logit Probability Matrix

Home	Away	Sample A Matrix =											LHS Vector =		Y
		X0	X1	X2	X3	X4	X5	X6	X7	X8	X9	X10	Probability	Ratio	Log (Ratio)
X1	X2	1	1	−1	0	0	0	0	0	0	0	0	.947	17.93	2.89
X1	X3	1	1	0	−1	0	0	0	0	0	0	0	.340	0.51	−0.66
X1	X4	1	1	0	0	−1	0	0	0	0	0	0	.860	6.13	1.81
X1	X5	1	1	0	0	0	−1	0	0	0	0	0	.512	1.05	0.05
X1	X6	1	1	0	0	0	0	−1	0	0	0	0	.639	1.77	0.57
⋯	⋯	⋯	⋯	⋯	⋯	⋯	⋯	⋯	⋯	⋯	⋯	⋯	⋯	⋯	⋯
Budget		1	1	1	1	1	1	1	1	1	1	1			25

Table 3.6B Logit Probability Model Results		
Team	Rating	Rank
1	2.019	8
2	− 0.163	10
3	3.042	4
4	0.575	9
5	2.664	5
6	2.057	7
7	3.436	3
8	3.969	2
9	2.531	6
10	4.647	1
HFA	0.222	

of the models in use fail at some point and provide very inaccurate out-of-sample predictions. Incorrect mathematics and improper modeling techniques will lead to erroneous conclusions. (But luckily for our readers, we have solved this important issue. You are welcome ☺).

If you are unfamiliar with matrix algebra or computing the rank of a matrix, please do not be discouraged at this point. The important issue is how to solve for the reduced rank problem. And the solution for solving for a reduced rank matrix is to introduce a budget constraint. In this case, the budget constraint represents the sum of all parameter values. We can simply state that the sum of all parameter values needs to equal 25 as an example. The exact value for the budget constraint will vary by number of teams and number of games. But analysts can determine the best-fit model based on changing the budget constraint value.

The y-vector (LHS) for this example is also shown in Table 3.6A. Notice that the values of y are computed directly from the probability that the home team will win the game from Table 3.5A. Finally, notice that the very last value in the y-column is 25, the specified value of the budget constraint.

The parameter values consisting of each team's rating can now be determined from the linear regression model. This model represented in matrix form is:

$$y = Ab$$

The solution and corresponding statistics can be determined directly from techniques presented in Chapter 2, Regression Models, or from a software pack such as MATLAB.

Our solution for these parameter ratings is in Table 3.6B.

An important feature here is that once we have ratings for all teams we are able to compute the probability of a team winning across all pairs of teams. This will allow us to compute the probability of winning across two teams that did play each other and where we do not have a probability value. The data used in this example does provide a probability of winning across all pairs of teams, but we are usually not provided with such a luxury in practice.

How close is the estimated probability to the actual probability? To answer this question, we calculate the expected probability that the home team will win the game using the team ratings above and our regression equation. This is:

$$y = b_0 + b_H - b_A$$

where b_0 is the home-field advantage parameter, b_h is the home team rating, and b_A is the away team rating. Recall that this regression equation is actually estimating the log of the probability Wins-Ratio, that is:

$$y = \ln\left(\frac{p}{1-p}\right)$$

Therefore, after some algebra and solving for probability p we have:

$$p = \frac{e^y}{1 + e^y}$$

Alternatively, the probability can be determined directly from the logistic equation and using the regression rating parameters. This is as follows:

$$p = \frac{1}{1 + e^{-(b_0 + b_H - b_A)}}$$

Readers can verify that these two equations yield the same result. In fact, it is a good check to ensure there are no errors in any of the calculations.

Example 3.7

To evaluate how well the estimated probability fit the actual probability from the LHS of the equation we ran a regression analysis on these terms. Please note that this is the same type of analysis where we would compare the results of the actual y variable with the estimated variable $\hat{y}$. A graphical illustration of this relationship is shown in Fig. 3.5.

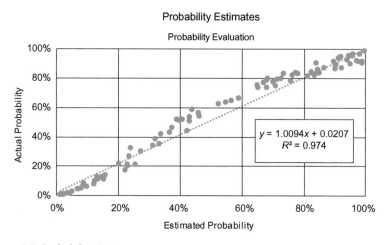

Figure 3.5 Probability Estimates.

Table 3.7 Probability Regression Results		
	Est. Probability	**Const**
Beta	1.009	0.021
SE	0.018	0.011
t-Stat	57.449	1.890
SeY	0.0546	
R^2	0.974	
F	3300.409	

This regression is as follows:

$$Actual\ Probability(\text{LHS}) = \beta_0 + \beta_1 \cdot Estimated\ Probability + e$$

This regression had an extremely high goodness of fit with $R^2 = 0.974$.

The results are in Table 3.7 and a graph of the actual probabilities as a function of estimated probabilities is in Fig. 3.5.

Example 3.8

We can also use the estimated home team winning probability to estimate the home team's expected winning margin. This is determined via running a regression of actual spread on the estimated probability as follows:

$$\text{Spread} = \beta_0 + \beta_1 \cdot Estimated\ Probability + e$$

The R^2 goodness of fit of this example is 66% with significant *t*-Stat. Additionally, the regression has a low error term $SeY = 3.76$. Thus, these results on their own indicate a predictive model.

A scatter plot of the actual spreads as a function of home winning probability level is shown in Fig. 3.6 and the regression equation results are shown in Table 3.8. The regression results comparing estimated probability to winning spread also have a high fit with $R^2 = 0.693$. The regression error term is 3.5. This indicates our estimate of the winning spread will be within $+/-3.5$ points.

Notice that these actual spreads as a function of probability are what we expect, that is, the average spread is positive when probability is greater than 50% (indicating the home team is expected to win), and the average spread is below zero when the probability is less than 50% (indicating the home team is expected to lose). But this scatter of the data around the best-fit line begins to show that a linear relationship between spreads and win probability may not be the best estimator of spread. Therefore, analysts can improve the estimation of the spread by using different functions such as an s-curve instead of a straight line. Additionally, this may be an indication that the data is not normally distributed. Recall from the introduction to this chapter that these analyses do require the data to be normally distributed.

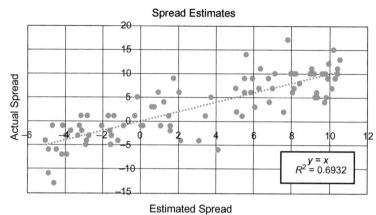

Figure 3.6 Spread Estimates.

	Table 3.8 Spread Probability Regression Results	
	Est. Probability	**Const**
Beta	15.851	−5.237
SE	1.124	0.702
t-Stat	14.101	−7.464
SeY	3.490	
R^2	0.693	
F	198.852	

It is important to note here that we may be realizing these results because:

1. The data is not normally distributed.
2. We need a different mapping from win probability to spreads to determine the best-fit equation, such as an s-curve.
3. In all cases we can incorporate an error correction term to adjust the data.
4. The data sample is too small and we may need either a larger number of games or a better cross-sectional number of games (i.e., more teams playing one another than just playing teams in their league or conference).

Example 3.9

Our next example demonstrates a technique that can be used to rate and rank teams using the logit probability model and incorporating game scores. This will also help resolve the issues that may arise with a team only winning against weaker opponents and a team only losing but against the strongest opponents. The technique provides an objective and unbiased approach to rank these teams based on the scores of the games, namely, the home team score minus the away team score. We will refer to this difference as the home team spread or home team margin, and in many cases, just as the spread or margin. Here, a positive value indicates that the home team won the game (i.e., the away team lost the game) and a negative value indicates that the home team lost the game (i.e., the away team won the game).

This example will also use the data provided in Table 3.5A. It is important to note that in this formulation, the construction of the A matrix is exactly the same. But the construction of the y (LHS) vector is based on the home team winning margin, denoted simply as "margin" or "spread."

The calculation of the *y*-value in this example is as follows:

Step 1. Compute the *Spread* as:

$$Spread = Home\ Team\ Score - Away\ Team\ Score$$

Step 2. Compute the *z*-score of the Spread "*z*" as:

$$z = \frac{Spread - avg(Spread)}{stdev(Spread)}$$

Notice that in this calculation we are computing a normalized spread since we are subtracting out the mean and dividing by its standard deviation. This *z*-score provides the number of standard deviations the actual spread was from its mean. Here, any value above zero indicates the home team wins and any value below zero indicates the away team wins.

Step 3. Determine the cumulative probability of *z*:

$$F(z) = NormsDist(z)$$

We now have a value $F(z)$ that only has values between 0 and 1 thus making a very good candidate for the logit regression. (Please note that we are using the normal distribution here to work through the example but analysts will need to determine the proper distribution of spreads to use in this calculation, e.g., normal, chi square, log-normal, etc.).

Step 4. Calculate *Y* as follows:

$$Y = \log\left(\frac{F(x)}{1 - F(x)}\right)$$

Notice that this transformation follows the exact same procedure as above using probabilities with the Wins-Ratio. An important point here is that we can make these transformations based on the *z*-score calculated from many different data sets such as spreads, home team points scored, away team points scored, and total points scored because the value of $F(x)$ will always be between 0 and 1. This approach will also be used in later chapters where we predict the number of points that a player is likely to achieve in a fantasy sports competition.

The sample y-vector (LHS) for this example is shown below for the first five games in Table 3.9A. We include the very last data point of 25 to represent the budget constraint value.

Readers will notice that the sample A matrix is the same matrix used in the logit probability model in Example 3.6. In the logit regressions this will always be the case.

Table 3.9A Sample Logit Spread Matrix

Home	Away		Sample A Matrix =											LHS Vector =		Y
		X0	X1	X2	X3	X4	X5	X6	X7	X8	X9	X10	Probability	Ratio	Log (Ratio)
X1	X2	1	1	−1	0	0	0	0	0	0	0	0	.673	2.06	0.72
X1	X3	1	1	0	−1	0	0	0	0	0	0	0	.363	0.57	−0.56
X1	X4	1	1	0	0	−1	0	0	0	0	0	0	.728	2.68	0.99
X1	X5	1	1	0	0	0	−1	0	0	0	0	0	.252	0.34	−1.09
X1	X6	1	1	0	0	0	0	−1	0	0	0	0	.363	0.57	−0.56
⋯	⋯	1	1	1	1	1	1	1	1	1	1	1	⋯	⋯	⋯
Budget															25

The only change that we can expect is that the LHS vector will have different values based on the data sets we are evaluating.

Additionally, it is important to point out here that the value of 25 in the budget constraint can be determined via statistical analysis. This value will vary based on the number of teams and games, as well as the parity across teams. Readers can experiment with different budget values to determine the value that works best with their specific sports model.

Again, the parameter values (e.g., each team's rating) can now be determined from the linear regression model. This model represented in matrix form is:

$$y = Ab$$

The solution and corresponding statistics can be determined directly from techniques presented in Chapter 2, Regression Models, or from a software pack such as MATLAB.

Our solution for these parameter ratings is shown in Table 3.9B.

We can now use this data to estimate the spread but our steps are in reverse order from 3.6 since we first estimate the spread, then winning probability.

The spread is computed as follows:

Step 1. Compute $F(z)$

$$F(z) = \frac{1}{1 + e^{-(b_0 + b_H - b_A)}}$$

Step 2. Convert $F(z)$ to spread using the inverse normal function and the average spread and standard deviation of the spread. These calculations are shown in Chapter 4, Advanced Math and Statistics.

$$Est.\ Spread = F^{-1}(z, avg(Spread), stdev(Spread))$$

For example, in the first game, we have home team #1 playing away team #2. Then we compute $F(z)$ using results in Table 3.9B as follows:

$$F(z) = \frac{1}{1 + e^{-(b_0 + b_H - b_A)}} = \frac{1}{1 + e^{-(-0.01 + 1.842 - 1.252)}} = 0.6412$$

Table 3.9B Logit Spread Model Results		
Team	**Rating**	**Rank**
1	1.842	8
2	1.252	9
3	3.144	3
4	0.773	10
5	2.133	7
6	2.444	5
7	3.103	4
8	4.256	1
9	2.281	6
10	3.782	2
HFA	-0.010	

Table 3.9C Logit Spread Probability Regression		
	Actual Spread	Const
Beta	0.980	0.097
SE	0.055	0.356
t-Stat	17.717	0.272
SeY	2.949	
R^2	0.781	
F	313.875	
Est. spread = 5.45		
SeY = 2.949		
Probability = 0.967705		

Then the estimated spread is:

$$Est.\ Spread = F^{-1}(0.6412, 3.19, 6.27) = 5.45$$

The winning probability can now be computed from the estimated spread as follows:

Step 1. Run regression of actual spread on estimated spread

$$Actual\ Spread = b_0 + b_1 \cdot Est.\ Spread + e$$

Step 2. Compute probability that the actual spread will be greater than zero using the regression results and regression error. That is:

$$Prob(Home > Away) = 1 - NormCDF(0,\ Est\ Spread,\ SeY)$$

The results of our regression are shown in Table 3.9C. This regression has a high fit with $R^2 = 0.78$ and a regression error of 2.949.

The probability that home team #1 will beat away team #2 is then calculated using the normal CDF (similar to its use in chapter: Regression Models). This is as follows:

$$Prob(Team\ 1 > Team\ 2) = 1 - NormCdf(0,\ 5.45,\ 2.949) = .9677$$

3.6 CONCLUSIONS

This chapter provided readers with an overview of various probability models that can be used to predict the winner of a game or match, rank teams, and estimate the winning margin, and compute the probability that a team will win.

These probability models consist of deriving "team rating" metrics that are subsequently used as the basis for output predictions. The techniques discussed include logistic and power function optimization models, logit regression models using win/loss observations, and cumulative spread distribution functions (CDFs). These models will serve as the foundation behind our sports prediction models discussed in later chapters.

REFERENCES

Enders, W. (1995). *Applied Econometric Time Series*. New York: John Wiley & Sons.
Greene, W. (2000). *Econometric Analysis* (4th ed.). Upper Saddle River, NJ: Prentice-Hall, Inc.
Hamilton, J. D. (1994). *Time Series Analysis*. Princeton, NJ: Princeton University Press.
Kennedy, P. (1998). *A Guide to Econometrics* (4th ed.). Cambridge, MA: The MIT Press.

CHAPTER 4

Advanced Math and Statistics

4.1 INTRODUCTION

In this chapter we provide an overview of probability and statistics and their use in sports modeling applications. The chapter begins with an overview of the mathematics required for probability and statistics modeling and a review of essential probability distribution functions required for model construction and parameter estimation. The chapter concludes with an introduction to different sampling techniques that can be used to test the accuracy of sports prediction models and to correct for data limitation problems. These data limitation issues are often present in sports modeling problems due to limited data observations and/or not having games across all pairs of teams.

The sampling techniques include with and without replacement, Monte Carlo techniques, bootstrapping, and jackknife techniques. These techniques are useful for sports such as soccer, basketball, football, and hockey, as well as baseball when we are evaluating the best mix of players to use and best lineup based on the opposing team's starting pitcher.

Finally, these techniques serve as the building blocks for the advanced applications and sports models that are discussed in Chapter 12, Fantasy Sports Models and Chapter 13, Advanced Modeling Techniques. A summary of these important and essential techniques include:

- Probability and Statistics
- Probability Distribution Function (PDF) and Cumulative Distribution Function (CDF)
- Sampling Techniques
  - With Replacement
  - Without Replacement
  - Monte Carlo Distribution
  - Bootstrapping
  - Jackknife Sampling

Optimal Sports Math, Statistics, and Fantasy.
DOI: http://dx.doi.org/10.1016/B978-0-12-805163-4.00004-9

4.2 PROBABILITY AND STATISTICS

A *random variable* is defined as a variable that can take on different values. These values are determined from its underlying probability distribution, and the actual distribution is characterized by a mean and standard deviation term (such as a normal distribution) also a skewness and a kurtosis measure. The value of the random variable is also often subject to random variations due to noise or chance.

A random variable can represent many different items such as expected daily temperature at a location in the middle of July, the expected attendance at a sporting event, a sports team's strength rating, as well as the probability that a team will win a game or score a specified number of points.

A random variable can also be the parameter of a model used to predict the outcome of the sports game. The goal of the analyst in this case is to compute an accurate estimate of this random variable parameter.

Random variables can be either discrete or continuous values. A discrete random variable can take on only a specific finite value or a countable list of values. For example, a discrete random variable in sports is the number of points that a team scores or the number difference between the home team points scored and away team points scored. A continuous random variable can take on any numerical value in an interval (and theoretically, have an infinite number of decimal places). For example, a continuous random variable in sports could be the team's strength rating or a performance metric such as batting average (which can both have an infinite number of decimals).

Probability Distributions

Mathematicians utilize *probability distribution* functions in many different ways. For example, probability distribution functions can be used to "quantify" and "describe" random variables, they can be used to determine statistical significance of estimated parameter values, they can be used to predict the likelihood of a specified outcome, and also to calculate the likelihood that an outcome falls within a specified interval (i.e., confidence intervals). As mentioned, these probability distribution functions are described by their mean, variance, skewness, and kurtosis terms.

A *probability mass function* (pmf) is a function used to describe the probability associated with the discrete variable. A *cumulative mass function* (cmf) is a function used to determine the probability that the observation will be less than or equal to some specified value.

In general terms, if x is a discrete random variable and x^* is a specified value, then the pmf and cmf functions are defined as follows:

Probability Mass Function (pmf):

$$f(x) = Prob(x = x^*)$$

Cumulative Mass Function (cmf):

$$F(x) = Prob(x \leq x^*)$$

Probability distribution functions for continuous random variables are similar to those for discrete random variables with one exception. Since the continuous random variable can take on any value in an interval the probability that the random variable will be equal to a specified value is thus zero. Therefore, the probability distribution function (pdf) for a continuous random variable defines the probability that the variable will be within a specified interval (say between a and b) and the cumulative distribution function for a continuous random variable is the probability that the variable will be less than or equal to a specified value x^*.

A *probability distribution function* (pdf) is used to describe the probability that a continuous random variable and will fall within a specified range. In theory, the probability that a continuous value can be a specified value is zero because there are an infinite number of values for the continuous random value. The *cumulative distribution function* (cdf) is a function used to determine the probability that the random value will be less than or equal to some specified value. In general terms, these functions are:

Probability Distribution Function (pdf):

$$Prob(a \leq X \leq b) = \int_a^b f(x)dx$$

Cumulative Distribution Function (cdf):

$$F(x) = Prob(X \leq x) = \int_{-\infty}^x f(x)dx$$

Going forward, we will use the terminology "pdf" to refer to probability distribution function and probability mass function, and we will use the terminology "cdf" to refer to cumulative distribution function and cumulative mass function.

Example 4.1 Discrete Probability Distribution Function

Consider a scenario where a person rolls two dice (die) and adds up the numbers rolled. Since the numbers on dice range from 1 to 6, the set of possible outcomes is from 2 to 12. A pdf can be used to show the probability of realizing any value from 2 to 12 and the cdf can be used to show the probability that the sum will be less than or equal to a specified value.

Table 4.1 shows the set of possible outcomes along with the number of ways of achieving the outcome value, the probability of achieving each outcome value (pdf), and the probability that the outcome value will be less than or equal to the outcome value (cdf). For example, there were 6 different ways to roll a 7 from two dice. These

Table 4.1 Discrete Random Variable: Rolling Die			
Value	Count	Pdf	Cdf
2	1	3%	3%
3	2	6%	8%
4	3	8%	17%
5	4	11%	28%
6	5	14%	42%
7	6	17%	58%
8	5	14%	72%
9	4	11%	83%
10	3	8%	92%
11	2	6%	97%
12	1	3%	100%
Total	36	100%	

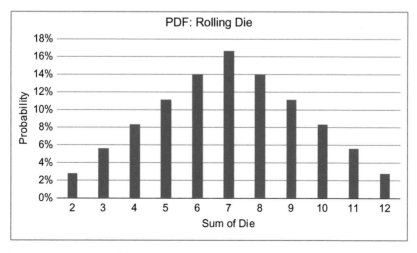

Figure 4.1 PDF: Rolling Die.

combinations are (1,6), (2,5), (3,4), (4,3), (5,2), and (6,1). Since there are 36 different combinations of outcomes from the two die, the probability of rolling a seven is 6/36 = 1/6, and thus, the pdf of 7 is 16.7%. Additionally, there are 21 ways that we can roll our die and have a value that is less than or equal to 7. Thus, the cdf is 21/36 = 58%. The pdf and cdf graphs for this example are shown in Figs. 4.1 and 4.2 respectively.

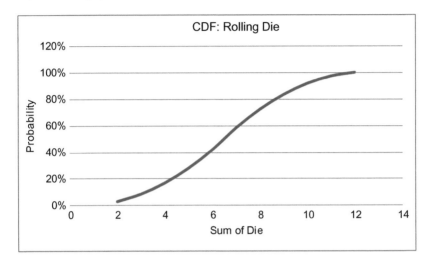

Figure 4.2 CDF: Rolling Die.

Example 4.2 Continuous probability distribution function

An example of a continuous probability distribution function can be best shown via the familiar standard normal distribution. This distribution is also commonly referred to as the Gaussian distribution as well as the bell curve.

Table 4.2 provides a sample of data for a standard normal distribution. The left-hand side of the table has the interval values a and b. The corresponding probability to the immediate right in this table shows the probability that the standard normal distribution

Table 4.2 Standard Normal Distribution				
a	b	Pdf	z	Cdf
−1	1	68.3%	−3	0.1%
−2	2	95.4%	−2	2.3%
−3	3	99.7%	−1	15.9%
−inf	−1	15.9%	0	50.0%
−inf	−2	2.3%	1	84.1%
1	inf	15.9%	2	97.7%
2	inf	2.3%	3	99.9%

will have a value between *a* and *b*. That is, if *x* is a standard normal variable, the probability that *x* will have a value between *a* and *b* is shown in the probability column.

For a standard normal distribution, the values shown in column "*a*" and column "*b*" can also be thought of as the number of standard deviations where 1 = plus one standard deviation and −1 = minus one standard deviation (and the same for the other values). Readers familiar with probability and statistics will surely recall that the probability that a standard normal random variable will be between −1 and +1 is 68.3%, the probability that a standard normal variable will be between −2 and +2 is 95.4%, and the probability that a standard normal variable will be between −3 and +3 is 99.7%.

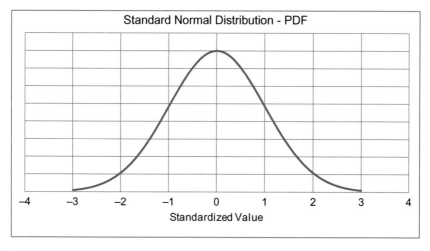

Figure 4.3 Standard Normal Distribution: PDF.

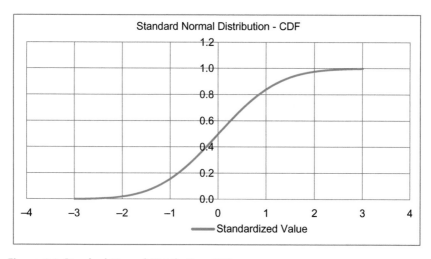

Figure 4.4 Standard Normal Distribution: CDF.

The data on the right-hand side of the table corresponds to the probability that a standard normal random value will be less than the value indicated in the column titled z. Readers familiar with probability and statistics will recall that the probability that a normal standard variable will be less than 0 is 50%, less than 1 is 84%, less than 2 is 97.7%, and less than 3 is 99.9%.

Fig. 4.3 illustrates a standard normal pdf distribution curve and Fig. 4.4 illustrates a standard normal cdf distribution curve. Analysts can use the pdf curves to determine the probability that an outcome event will be within a specified range and can use the cdf curves to determine the probability that an outcome event will be less than or equal to a specified value. For example, we utilize these curves to estimate the probability that a team will win a game and/or win a game by more than a specified number of points. These techniques are discussed in the subsequent sports chapters.

Important Notes:
- One of the most important items regarding computing probabilities such as the likelihood of scoring a specified number of points, winning a game, or winning by at least a specified number of points is using the proper distribution function to compute these probabilities.
- Different distribution functions will have different corresponding probability values for the same outcome value.
- It is essential that analysts perform a thorough review of the outcome variable they are looking to estimate and determine the correct underlying distribution.
- While there are many techniques that can be used to determine the proper distribution functions, analysts can gain important insight using histograms, p-p plots, and q-q plots as the starting points.
- We provide information about some of the more useful distributions below and analysts are encouraged to evaluate a full array of these distributions to determine which is most appropriate before drawing conclusions about outcomes, winning teams, scores, etc.

Descriptive Statistics

Each probability distribution has a set of descriptive statistics that can be used in analysis. The more important descriptive statistics for sports models are:

Mean: The arithmetic mean, also known as the simple mean or equal weighted mean. The mean of a data series is a unique value. The mean is also known as the first moment of the data distribution.

$$\mu = \frac{1}{n}\sum_{i=1}^{n} x_i$$

Mode: The value(s) of a data series that occurs most often. The mode of a data series is not a unique value.

Median: The value of a data series such that one-half of the observations are lower or equal and one-half the observations are higher or equal value. The median value is not a unique number. For example, in the series 1, 2, 3 the median is the value 2. But in the series 1, 2, 3, 4 there is not a unique value. Any number $2 < x < 3$ is the median of this series since exactly 50% of the data values are lower than x and exactly 50% of the data points are higher than x. A general rule of thumb is that if there are an odd number of data points, the middle value is the median, and if there is an even number of data points, the median is selected as the mean of the two middle points. In our example, 1, 2, 3, 4 the median would be taken as 2.5. However, any value x such that $2 < x < 3$ would also be correct.

Standard Deviation: The amount of dispersion around the mean. A small standard deviation indicates that the data are all close to the mean and a high standard deviation indicates that the data could be far from the mean. The standard deviation $\sigma(x)$ is the square root of the variance $V[x]$ of the data. The variance is also known as the second moment about the distribution mean.

$$\sigma^2 = \frac{1}{n}\sum_{i=1}^{n}(x-\mu)^2$$

$$\sigma = \sqrt{\sigma^2} = \sqrt{\frac{1}{n}\sum_{i=1}^{n}(x-\mu)^2}$$

Coefficient of Variation: A measure of the standard deviation divided by the mean. The coefficient of variation serves as a normalization of the data for a fair comparison of data dispersion across different values (e.g., as a measure of data dispersion of daily or monthly stock trading volumes).

$$COV = \frac{\sigma}{\bar{x}}$$

Skewness: A measure of the symmetry of the data distribution. A positively skewed data distribution indicates that the distribution has more data on the right tail (data is positively skewed). A negatively skewed data distribution indicates that the distribution has more data on the left tail (data is negatively skewed). A skewness measure of zero indicates that the data is symmetric. Skewness is also known as the third moment about the mean.

$$\text{Skewness} = \sqrt{\frac{1}{n} \sum_{i=1}^{n} \frac{(x - \mu)^3}{\sigma}}$$

Kurtosis: A measure of the peakedness of the data distribution. Data distributions with negative kurtosis are called platykurtic distributions and data distributions with positive kurtosis are called leptokurtic distributions.

$$\text{Kurtosis} = \sqrt{\frac{1}{n} \sum_{i=1}^{n} \frac{(x - \mu)^3}{\sigma^2}}$$

Probability Distribution Functions

In this section we provide a description of the important probability distribution functions that are used in sports modeling. Readers interested in a more thorough investigation of these distributions are referred to Meyer (1970), Dudewicz and Mishra (1988), Pfeiffer (1978), DeGroot (1986).

Our summary table of the distribution statistics and moments is based on and can also be found at: www.mathworld.wolfram. com, www.wikipedia.org/, www.statsoft.com/textbook/, and www. mathwave.com/articles/distribution_fitting.html. These are excellent references and are continuously being updated with practical examples. These probability and distribution functions below are also a subset of those presented in Glantz and Kissell (2013) and used for financial risk modeling estimation.

Continuous Distribution Functions

Normal Distribution

A normal distribution is the workhorse of statistical analysis. It is also known as the Gaussian distribution and the bell curve (for the distribution's resemblance to a bell). It is one of the most used distributions in statistics and is used for several different applications. The normal distribution also provides insight into issues where the data is not necessarily normal, but can be approximated by a normal distribution. Additionally, by the central limit theorem of mathematics we find that the mean of a sufficiently large number of data points will be normally distributed. This is extremely useful for parameter estimation analysis such as with our regression models.

Normal Distribution Statistics[1]	
Notation	$N(\mu, \sigma^2)$
Parameter	$-\infty < \mu < \infty$ $\sigma^2 > 0$
Distribution	$-\infty < x < \infty$
Pdf	$\dfrac{1}{\sqrt{2\pi\sigma}} \exp\left\{-\dfrac{(x-\mu)^2}{2\sigma^2}\right\}$
Cdf	$\dfrac{1}{2}\left[1 + \operatorname{erf}\left(\dfrac{x-\mu}{2\sigma^2}\right)\right]$
Mean	μ
Variance	σ^2
Skewness	0
Kurtosis	0

where erf is the Gauss error function, i.e.,

$$\operatorname{erf}(x) = \frac{2}{\sqrt{\pi}} \int_0^x \exp(-t^2)$$

Normal Distribution Graph

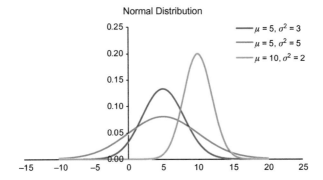

Standard Normal Distribution

The standard normal distribution is a special case of the normal distribution where $\mu = 0$, $\sigma^2 = 1$. If is often essential to normalize data prior to the analysis. A random normal variable with mean μ and standard deviation μ can be normalized via the following:

$$z = \frac{x - \mu}{\sigma}$$

Standard Normal Distribution Statistics[1]	
Notation	$N(0, 1)$
Parameter	n/a
Distribution	$-\infty < z < \infty$
Pdf	$\dfrac{1}{\sqrt{2\pi}}\exp\left\{-\dfrac{1}{2}z^2\right\}$
Cdf	$\dfrac{1}{2}\left[1 + \text{erf}\left(\dfrac{z}{2}\right)\right]$
Mean	0
Variance	1
Skewness	0
Kurtosis	0

Standard Normal Distribution Graph

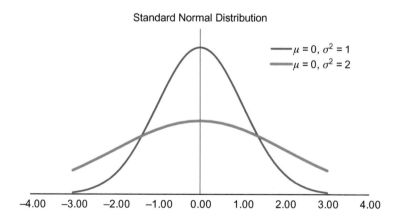

Standard Normal Distribution

Student's t-Distribution

Student's t-distribution (a.k.a. t-distribution) is used when we are estimating the mean of normally distributed random variables where the sample size is small and the standard deviation is unknown. It is used to perform hypothesis testing around the data to determine if the data is within a specified range. The t-distribution is used in hypothesis testing of regression parameters (e.g., when developing risk factor models). The t-distribution looks very similar to the normal distribution but with fatter tails. But it also converges to the normal curve as the sample size increases.

Student's t-Distribution[1]	
Notation	$t\text{-dist}(\nu)$
Parameter	$\nu > 0$
Distribution	$-\infty < x < \infty$
Pdf	$\dfrac{\Gamma\left(\frac{\nu+1}{2}\right)}{\sqrt{\nu\pi}\,\Gamma\left(\frac{\nu}{2}\right)}\left(1+\dfrac{x^2}{\nu}\right)^{-\frac{\nu+1}{2}}$
Cdf	
Mean	$= \begin{cases} 0 & \nu > 1 \\ undefined & o.w. \end{cases}$
Variance	$= \begin{cases} \dfrac{\nu}{\nu+1} & \nu > 2 \\ \infty & 1 < \nu \leq 2 \\ undefined & o.w. \end{cases}$
Skewness	$= \begin{cases} 0 & \nu > 3 \\ undefined & o.w. \end{cases}$
Kurtosis	$= \begin{cases} \dfrac{6}{\nu-4} & \nu > 4 \\ \infty & 2 < \nu \leq 4 \\ undefined & o.w. \end{cases}$

Student's *t*-Distribution Graph

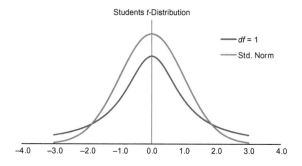

Student's *t*-Distribution: Interesting Notes

Have you ever wondered why many analysts state that you need to have at least 20 data points to compute statistics such as average or standard deviation? The reason is that once there are 20 data points, Student's *t*-distribution converges to a normal distribution. Then analysts could begin to use the simpler distribution function.

Where did the name "Student's t-distribution" come from? In many academic textbook examples, the Student's *t*-distribution is used to estimate their performance from class tests (e.g., midterms and finals, standardized tests, etc.). Therefore, the *t*-distribution is the appropriate distribution since it is a small sample size and the standard deviation is unknown. But the distribution did not arise from evaluating test scores. The Student's *t*-distribution was introduced to the world by William Sealy Gosset in 1908. The story behind the naming of the Student's *t*-distribution is as follows: William was working at the Guinness Beer Brewery in Ireland and published a paper on the quality control process they were using for their brewing process. And to keep their competitors from learning their processing secrets, Gosset published the test procedure he was using under the pseudonym Student. Hence, the name of the distribution was born.

Student's Distribution Graph
(with *k* = 10, 20, 100 and normal curve)

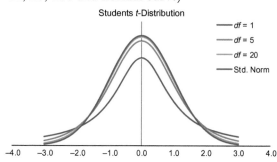

Log-Normal Distribution

A log–normal distribution is a continuous distribution of random variable y whose natural logarithm is normally distributed. For example, if random variable $y = \exp\{y\}$ has log-normal distribution then $x = \log(y)$ has normal distribution. Log-normal distributions are most often used in finance to model stock prices, index values, asset returns, as well as exchange rates, derivatives, etc.

Log-Normal Distribution Statistics[1]	
Notation	$\ln N(\mu, \sigma^2)$
Parameter	$-\infty < \mu < \infty$ $\sigma^2 > 0$
Distribution	$x > 0$
Pdf	$\dfrac{1}{\sqrt{2\pi}\sigma x} \exp\left\{ -\dfrac{(\ln(x)-\mu)^2}{2\sigma^2} \right\}$
Cdf	$\dfrac{1}{2}\left[1 + \operatorname{erf}\left(\dfrac{\ln(x-\mu)}{\sigma} \right) \right]$
Mean	$e^{\left(\mu + \frac{1}{2}\sigma^2\right)}$
Variance	$(e^{\sigma^2} - 1)e^{2\mu + \sigma^2}$
Skewness	$(e^{\sigma^2} + 2)\sqrt{(e^{\sigma^2} - 1)}$
Kurtosis	$e^{4\sigma^2} + 2e^{3\sigma^2} + 3e^{2\sigma^2} - 6$

where erf is the Gaussian error function.

Log-Normal Distribution Graph

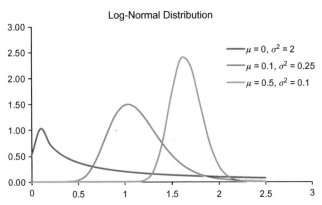

Uniform Distribution

The uniform distribution is used when each outcome has the same likelihood of occurring. One of the most illustrated examples of the uniform distribution is rolling a die where each of the six numbers has equal likelihood of occurring, or a roulette wheel where (again) each number has an equal likelihood of occurring. The uniform distribution has constant probability across all values. It can be either a discrete or continuous distribution.

Uniform Distribution Statistics[1]	
Notation	$U(a, b)$
Parameter	$-\infty < a < b < \infty$
Distribution	$a < x < b$
Pdf	$\dfrac{1}{b - a}$
Cdf	$\dfrac{x - a}{b - a}$
Mean	$\dfrac{1}{2}(a + b)$
Variance	$\dfrac{1}{12}(b - a)^2$
Skewness	0
Kurtosis	$-\dfrac{6}{5}$

Uniform Distribution Graph

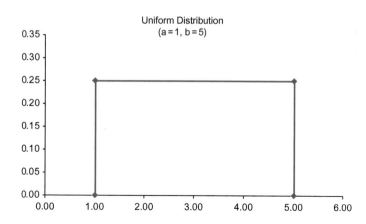

Exponential Distribution

The exponential distribution is a continuous distribution that is commonly used to measure the expected time for an event to occur. For example, in physics it is often used to measure radioactive decay, in engineering it is used to measure the time associated with receiving a defective part on an assembly line, and in finance it is often used to measure the likelihood of the next default for a portfolio of financial assets. It can also be used to measure the likelihood of incurring a specified number of defaults within a specified time period.

Exponential Distribution Statistics[1]	
Notation	$Exponential(\lambda)$
Parameter	$\lambda > 0$
Distribution	$x > 0$
Pdf	$\lambda e^{-\lambda x}$
Cdf	$1 - e^{-\lambda x}$
Mean	$1/\lambda$
Variance	$1/\lambda^2$
Skewness	2
Kurtosis	6

Exponential Distribution Graph

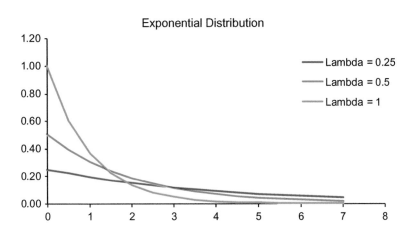

Exponential Distribution

Chi-Square Distribution

A chi-square distribution is a continuous distribution with k degrees of freedom. It is used to describe the distribution of a sum of squared random variables. It is also used to test the goodness of fit of a distribution of data, whether data series are independent, and for estimating confidences surrounding variance and standard deviation for a random variable from a normal distribution. Additionally, chi-square distribution is a special case of the gamma distribution.

Chi-Square Distribution Statistics[1]	
Notation	$\chi(k)$
Parameter	$k = 1, 2, \ldots$
Distribution	$x \geq 0$
Pdf	$\left(x^{\frac{k}{2}-1} e^{-\frac{x}{2}} \right) / \left(2^{\frac{k}{2}} \Gamma\left(\frac{k}{2}\right) \right)$
Cdf	$\gamma\left(\frac{k}{2}, \frac{x}{2}\right) / \Gamma\left(\frac{k}{2}\right)$
Mean	k
Variance	$2k$
Skewness	$\sqrt{8/k}$
Kurtosis	$12/k$

where $\gamma\left(\frac{k}{2}, \frac{x}{2}\right)$ is known as the incomplete Gamma function (www.mathworld.wolfram.com).

Chi-Square Distribution Graph

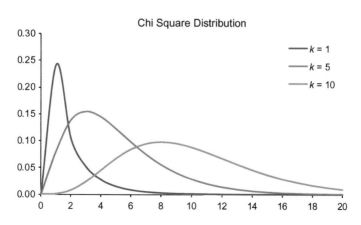

Logistic Distribution

The logistic distribution is a continuous distribution function. Both its pdf and cdf functions have been used in many different areas such as logistic regression, logit models, neural networks. It has been used in the physical sciences, sports modeling, and recently in finance. The logistic distribution has wider tails than a normal distribution so it is more consistent with the underlying data and provides better insight into the likelihood of extreme events.

Logistic Distribution Statistics[1]	
Notation	$Logistic(\mu, s)$
Parameter	$0 \leq \mu \leq \infty$ $s > 0$
Distribution	$0 \leq x \leq \infty$
Pdf	$\dfrac{\exp\left(-\frac{x-\mu}{s}\right)}{s\left(1+\exp\left(-\frac{x-\mu}{s}\right)\right)^{2}}$
Cdf	$\dfrac{1}{1+\exp\left(-\frac{x-\mu}{s}\right)}$
Mean	μ
Variance	$\dfrac{1}{3}s^{2}\pi^{2}$
Skewness	0
Kurtosis	$6/5$

Logistic Distribution Graph

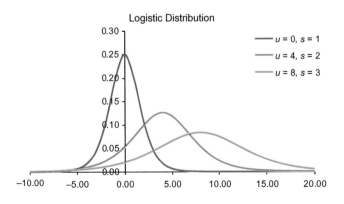

Triangular Distribution

The triangular distribution is when there is a known relationship between the variable data but when there is relatively little data available to conduct a full statistical analysis. It is often used in simulations when there is very little known about the data-generating process and is often referred to as a "lack of knowledge" distribution. The triangular distribution is an ideal distribution when the only data on hand are the maximum and minimum values, and the most likely outcome. It is often used in business decision analysis.

Triangular Distribution[1]	
Notation	$\text{Triangular}(a, b, c)$
Parameter	$-\infty \leq a \leq \infty$ $b > a$ $a < c < b$
Distribution	$a < x < b$
Pdf	$= \begin{cases} \dfrac{2(x-a)}{(b-a)(c-a)} & a \leq x \leq c \\ \dfrac{2(x-a)}{(b-a)(b-c)} & c \leq x \leq b \end{cases}$
Cdf	$= \begin{cases} \dfrac{2(x-a)^2}{(b-a)(c-a)} & a \leq x \leq c \\ 1 - \dfrac{(b-x)^2}{(b-a)(b-c)} & c \leq x \leq b \end{cases}$
Mean	$\dfrac{a+b+c}{3}$
Variance	$\dfrac{a^2 + b^2 + c^2 - ab - ac - bc}{18}$
Skewness	$\dfrac{\sqrt{2}(a+b-2c)(2a-b-c)(a-2b+c)}{5(a^2+b^2+c^2-ab-ac-bc)^{\frac{3}{2}}}$
Kurtosis	$-\dfrac{3}{5}$

Triangular Distribution Graph

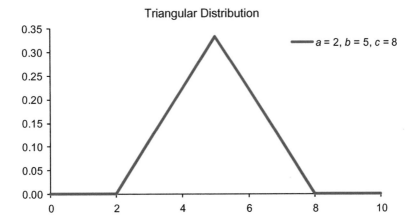

Discrete Distributions

Binomial Distribution

The binomial distribution is a discrete distribution used for sampling experiments with replacement. In this scenario, the likelihood of an element being selected remains constant throughout the data-generating process. This is an important distribution in finance in situations where analysts are looking to model the behavior of the market participants who enter reserve orders to the market. Reserve orders are orders that will instantaneously replace if the shares are transacted. For example, if an investor who has 1000 shares to buy entered at the bid may be showing 100 shares to the market at a time. Once those shares are transacted the order immediately replenishes (but the priority of the order moves to the end of the queue at that trading destination at that price). These order replenishments could occur with a reserve or iceberg type of order or via high-frequency trading algorithms where once a transaction takes place the market participant immediately submits another order at the same price and order size thus giving the impression that the order was immediately replaced.

Binomial Distribution Statistics[1]	
Notation	$\text{Binomial}(n, p)$
Parameter	$n \geq 0\ 0 \leq p \leq 1$
Distribution	$k = 1, 2, \ldots, n$
Pdf	$\binom{n}{k} p^k (1-p)^{n-k}$
Cdf	$\sum_{i=1}^{k} \binom{n}{i} p^i (1-p)^{n-i}$
Mean	np
Variance	$np(1 - p)$
Skewness	$\dfrac{1 - 2p}{\sqrt{np(1 - p)}}$
Kurtosis	$\dfrac{1 - 6p(1 - p)}{np(1 - p)}$

Binomial Distribution Graph

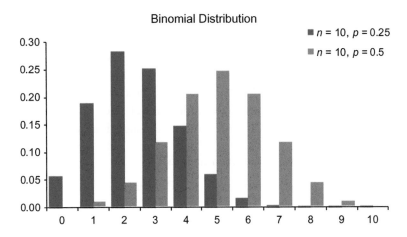

Poisson Distribution

The Poisson distribution is a discrete distribution that measures the probability of a given number of events happening in a specified time period. In finance, the Poisson distribution could be used to model the arrival of new buy or sell orders entered into the market or the expected arrival of orders at specified trading venues or dark pools. In these cases, the Poisson distribution is used to provide expectations surrounding confidence bounds around the expected order arrival rates. Poisson distributions are very useful for smart order routers and algorithmic trading.

Poisson Distribution Statistics[1]	
Notation	$\text{Poisson}(\lambda)$
Parameter	$\lambda > 0$
Distribution	$k = 1, 2, \ldots,$
Pdf	$\dfrac{\lambda^k e^{-\lambda}}{k!}$
Cdf	$\displaystyle\sum_{i=1}^{k} \dfrac{\lambda^k e^{-\lambda}}{k!}$
Mean	λ
Variance	λ
Skewness	$\lambda^{-1/2}$
Kurtosis	λ^{-1}

Poisson Distribution Graph

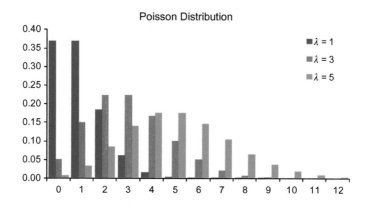

4.3 SAMPLING TECHNIQUES

What is data sampling? Data sampling is a statistical technique that is used to ascertain information about an outcome event such as a predicted score or probability of winning, or information about a specified model including the significance of the explanatory factors and parameters, or information about the underlying probability distribution using a subset of data rather than the entire data universe.

Data sampling is required when:

1. We are unable to observe and collect all data across all possible outcomes;
2. The collection of all data outcomes is not easily manageable;
3. We need to understand the accuracy of the model including significance of the parameters and distribution of the data;
4. We do not have a sufficient number of data points for a complete and thorough analysis.

For example, during a presidential election it is not possible to poll all voters to determine their favorite candidate and likely election winter. Thus, statisticians seek to draw conclusions about the likely winner using a smaller subset of data, known as a sample.

Furthermore, in sports modeling problems, we very often do not have enough observations across all teams and possible pairs of games to incorporate into our models. And in many types of sports competitions we do not have observations across all potential sets of teams to be able make a prediction based on actual data or historical data. This is the case with MLB, NFL, and college sports, as well as in international soccer or FIFA competitions, we do not have games or observations between all pairs of teams, thus making it difficult to draw conclusions. In all of these situations, we are left with making inferences and constructing models using a subset or limited amount of data.

Data sampling helps analysts resolve data limitation problems and generate outcome predictions. It allows modelers to utilize smaller data sets and/or incomplete data sets and build and test models efficiently. Data sampling, however, is associated with uncertainty and sampling error. It is required that the analyst understands the statistical error and uncertainty when making predictions about an upcoming game. As it turns out, understanding the statistical accuracy of the model and the underlying distribution of the error term is one of the most important functions of the data modeling process. In many situations, sampling of the data sample is needed to generate these error terms and to understand the distribution of these error terms. Many of the more important probability distribution functions for sports modeling problems are described above.

The remainder of this chapter will discuss different types of data sampling techniques and their use in sports modeling problems. These techniques include:

- Random Sampling
- Sampling with Replacement
- Sampling without Replacement
- Monte Carlo Techniques
- Bootstrapping Techniques
- Jackknife Sampling Techniques

4.4 RANDOM SAMPLING

Random sampling is a statistical technique that selects a data sample based upon a predefined probability that each data point may be selected for analysis. The probability levels are determined in a manner such that the underlying data subset will be most appropriate for the data modeling needs. In many cases these probability levels are specified such that each data point will have the same chance of being included and in other cases the probability levels are specified such that the expected data set will have consistent and/or similar characteristics as the data universe.

Nonrandom sampling is another sampling technique. In this case, the actual data samples are selected based on availability or ease of the data collection process. Data points are not selected based on any probability level, and thus, the likelihood of any data item being included in the subset sample will differ. This makes it difficult to make inferences about the larger data universe and introduces additional error into the modeling process. However, there are techniques that analysts can use to account for these biases. Many of these nonrandom sampling techniques are used in qualitative surveys where a surveyor stands at the front of a mall, supermarket, train station, or some other location and asks questions to people walking by. Thus, only the people who would be visiting these sites at these times could become part of the sample. These types of nonrandom sampling techniques include convenience sampling, consecutive sampling, and quota sampling techniques. These sampling techniques are not appropriate sampling techniques for sports modeling problems and will not be discussed in the text.

Resampling is a statistical technique that consists of performing an analysis, running a model, or estimating parameter values for many different data sets where these data sets are selected from the larger data universe. Resampling is an appropriate technique for many different statistical applications and can be used to estimate parameter values and probability distributions. In many situations, as mentioned above, we may not have enough data points or data observations to be able to use these metrics directly due to data limitation issues, and/or the underlying mathematical model may be too complex to calculate error terms due to data limitations.

Resampling allows analysts to estimate parameter values and probability distributions using the data samples. This then allows analysts to evaluate, test, and critique modeling approaches to determine the best and most appropriate model for problem. Resampling allows analysts to make proper statistical inferences and conclusions about future outcome events using only the data at hand.

4.5 SAMPLING WITH REPLACEMENT

Sampling with replacement is a resampling technique where each data item can be selected for and included in the data sample subset more than once. For example, suppose we have a bag of ping pong balls with numbers written on each ball. If we are interested in learning the average number written on the ping pong ball using a sampling with replacement approach, we would pick a ball at random, write down the number, and then put the ball back in the bag. Then we would pick another ball at random, write down the number, and then put the ball back in the bag. The selection of balls would be repeated for a specified number of times. Once completed, we would calculate the average across all numbers written down. In this analysis, it is quite possible to pick the same ball multiple times.

Sampling with replacement is similar to many lotto games where the player picks four numbers from 1 to 10 and where each number can be selected more than once. In this scenario, there would be 4 machines with 10 ping pong balls each numbered from 1 to 10. Then the machines would select one ball from each machine. The four numbers selected could consist of all different numbers such as 1-2-8-4 or have some or all repeated numbers such as 5-2-5-1 or 9-9-9-9.

- If a data item can be selected more than once it is considered sampling with replacement.

4.6 SAMPLING WITHOUT REPLACEMENT

Sampling without replacement is a resampling technique where each data item can be selected and used on our data sample subset only once. For example, using the same ping pong ball example where we are interested in learning the average value of the numbers on the ping pong balls the sampling without replacement would consist of picking a ball from the bag at random and writing down its value, but leaving the ball outside of the bag, and then picking another ball from the bag, writing down its value, and leaving that ball outside the bag, and repeating this process for a specified number of draws. In this case, each ball can only be selected one single time.

Sampling without replacement is similar to a Powerball type of contest where a player is asked to pick 6 numbers from 1 to 44 (or variations of this type of selection). In this scenario, each number can only be selected a single time.

- If a data item can only be selected one time than it is considered sampling without replacement.

4.7 BOOTSTRAPPING TECHNIQUES

Bootstrapping is a statistical technique that refers to random sampling of data with replacement. One of the main goals of bootstrapping is to allow analysts to estimate parameter values, corresponding standard errors, and to gain an understanding of the probability distribution of the model's error term.

In sports modeling problems, bootstrapping sampling techniques are essential for being able to calculate a statistically accurate team strength rating and also to be able to accurately predict the outcome of an event or game. This is especially true in situations where we may not have a large enough number of observations of games across all teams and/or situations where all teams may not play against each other during the season. Thus, it is important for all professional sports and college sports modeling problems.

For a bootstrapping sample, analysts could simply select a specified sample size, such as 25% of the actual data. Thus, if there are 1000 observations each sample could consist of 250 data points. Bootstrapping techniques use sampling with replacement so each data point can be selected for a sample more than one time. The model is then solved repeatedly. With today's computing power we can set the number of actual repeated samples to be quite large such as $N \gg 1000$ to allow accurate parameter estimates and confidence levels. That is, we can sample the data and solve the model 1000 times or more.

Consider the power function presented in Chapter 2, Regression Models, where we seek to maximize the following:

$$Max: \ \log L = \sum_{i=1}^{G} \ln \left(\frac{b_0 + b_h}{b_0 + b_h + b_a} \right)$$

where G is the total number of games in the sample and b_k represents the model parameters.

If we solve this optimization once we only have the parameter estimates (e.g., team strength rating) for each team and the home-field advantage term. But we do not have any estimates surrounding the standard errors of the parameters.

However, by performing bootstrapping sampling using say 25% of the games for each optimization solution and repeating this sampling technique 1000 times or more, we can calculate both the team rating parameter and the standard error of the parameter value, thus allowing us to statistically evaluate the model and mark comparisons across teams.

Using bootstrapping techniques, the expected parameters value is taken as the average value across all samples and the confidence interval or standard error can be computed using either the standard deviation of parameter estimates or computed from a specified middle percentile interval such as middle 50% or middle 68% (to be consistent with the standard deviation) of data points. It is important to note that using the standard deviation of results to compute standard errors in this case may be inaccurate in times of small sizes. Analysts will need to understand how the sample size affects the parameters estimates for their particular sports models or application.

4.8 JACKKNIFE SAMPLING TECHNIQUES

Jackknife sampling is another type of resampling technique that is used to estimate parameter values and corresponding standard deviations similar to bootstrapping. The sampling method for the jackknife technique requires that the analyst omit a single observation in each data sample. Thus, if there are n data points in the sample, the jackknife sampling technique will consist of n samples each with $n - 1$ data points in each sample subset analysis. Thus, in this case, the analyst would solve the model n times each with $n-1$ data point. This would allow the analyst to estimate both parameter value and corresponding standard error.

Our research into sports modeling problems, however, finds that this jackknife technique may yield inconsistent results across teams and parameters especially in situations where a team wins all of its games or a high percentage of games, in situations where a team loses all of its games or a high percentage of its games, and/or where a team plays both very strong and very weak opponents—which is very common across college sports and also in many international tournaments such as FIFA soccer and other World Cup tournaments.

An appropriate adjustment to the jackknife sampling technique in these situations is to entirely leave out a team and all of its games in each data sample. For example, if there are 100 games across 10 teams where each team played 10 games, our jackknife sampling technique would consist of 10 samples each with 90 games. So if team A played 10 games and we are leaving team A out of this same run we would omit the 10 records with team A. While both variations of the jackknife sampling techniques have advantages and disadvantages, analysts will need to determine from the data which is the most appropriate technique to use and in which types of situations it is most appropriate and accurate.

Therefore, if we are looking to estimate team rating parameter values using the power function described in Chapter 2, Regression Models, for a scenario with 25 teams, we would solve the following optimization problem 25 times and in each optimization sample we would leave out one team and all of its games.

Thus, we would maximize the following:

$$Max: \ \log L = \sum_{i=1}^{G_k} \ln\left(\frac{b_0 + b_h}{b_0 + b_h + b_a}\right)$$

where G_k consists of all the games that did not involve team K. If there are M teams in total, we would repeat this optimization M times. Here, b_k represents the model parameters. Additionally, it is important to note that each team will have $M-1$ parameter values, one value for each scenario that included their team.

The estimated parameter values, or in this case team rating values, are computed from the optimization results across all M samples. The expected parameter value is the average across all $M-1$ results for each team. The standard error term is computed as the middle percentile values such as the middle 50% values or middle 68% values (to be consistent with standard deviation). The standard error can also be computed as the standard deviation across all $M-1$ parameter estimates.

Again, it is important for analysts to understand the effect of small samples and data limitations on their model and error terms. Analysts need to investigate the actual model error term to determine which is the most appropriate technique to quantify standard errors.

Finally, once we determine team rating parameters and corresponding standard errors, we can use this information to make statistically accurate rankings and comparisons across teams, and predict outcome events with a high level of accuracy, e.g., predict the expected winner of a game, calculate the expected winning margin, compute the winning probability.

4.9 MONTE CARLO SIMULATION

Monte Carlo simulation is a statistical technique that predicts outcomes based on probability estimates and other specified input values. These input values are often assumed to have a certain distribution or can take on a specified set of values.

Monte Carlo simulation is based on repeatedly sampling the data and calculating outcome values from the model. In each sample, the input factor and model parameters can take on different values. These values are simulated based on the distribution of the input factor and parameter values. For example, if X is an input factor for our model and X is a standard normal random variable, each simulation will sample a value of X from a standard normal distribution. Thus, each sample scenario will have a different value of X. Analysts then run repeated simulations where they can allow both the parameter values and input factors to vary based on their mean and standard error. Analysts then use these the results of these simulations to learn about the system and make better-informed future decisions.

Another important use of Monte Carlo simulation is to evaluate the performance and accuracy of a model, and also to evaluate whether or not a model or modeling methodology is appropriate for certain situation. For example, we discussed in Chapter 2, Regression Models, the power function and how it can be used in sports modeling as the basis for predicting the winning team, the winning score, and probability of winning. We can use these same Monte Carlo simulation techniques to determine whether or not this technique is appropriate for the sport we are looking to model. The process is as follows:

Suppose we want to determine if the power function and optimization process is an appropriate modeling technique to rank college football teams and to predict the winning team. Here, we apply Monte Carlo simulation as follows:

Step 1: Assign each team a rating score b_i that indicates the team's overall strength. These ratings can be assigned via a completely random process, or they can be assigned based on the team's previous years' winning records, based on conferences, etc. We suggest running multiple trials where the ratings are assigned via various methods in order to best analyze the model. It is important to note here that once we assign the team rating score b_i to each team we then know the exact ranking of teams and the exact probability that any one team will beat any other team.

Step 2: Run a simulation of game outcomes based on an actual schedule of games. Similar to Step 1, we suggest repeating this experiment using the schedules from different years in order to fully test the process and specified model.

Step 3: Determine the winner of a game based on the team's rating and power function, and based on a simulated random value (between 0 and 1). For example, if home team A is playing away team B, the probability that home team A will win the game is determined from the power function as follows:

$$P(A > B) \frac{b_0 + b_A}{b_0 + b_A + b_B}$$

If x is the randomly generated value (with a value between 0 and 1), we assign team A as the winner of the game if $x \leq P$ and we assign team B as the winner of the game if $x > P$.

Step 4: Simulate the winner of every game on the schedule across all teams during a season based on Step 3.

Step 5: Solve for each team's estimated rating value based on the outcomes of the simulated season.

Step 6: Compare the rating results obtained from the simulated season to the actual rating values used to simulate the season results. If the model is appropriate for the sport we should find a high correlation between actual ratings and estimated ratings. And, the rankings of the teams from the simulated results should be consistent with the actual rankings used to simulate the results. If either of these is found to be inconsistent with actual values used in the simulation, then the model would not be appropriate for the sport in question.

Step 7: Repeat this simulation test for various scenarios where teams are given different rating values and using schedules from different seasons.

4.10 CONCLUSION

In this chapter we provided readers with an overview of the essential mathematics required for probability and statistics modeling. The chapter included insight into different probability distribution functions and the important mathematical metrics used to describe these functions. We also provided readers with an overview of different sampling techniques and how these techniques can be used to evaluate, test, and critique sports prediction models.

ENDNOTE

1. www.mathworld.wolfram.com/topics/ProbabilityandStatistics.html

REFERENCES

DeGroot, M. H. (1986). *Probability and Statistic* (2nd ed.). New York, NY: Addison Wesley.
Dudewicz, E., & Mishra, S. (1988). *Modern Mathematical Statistics*. New York, NY: John Wiley & Sons.
Glantz, M., & Kissell, R. (2013). *Multi-Asset Risk Modeling: Techniques for a global economy in an electronic and algorithmic trading era*. Elsevier.
Meyer, P. (1970). *Introductory Probability and Statistical Applications* (2nd ed.). Addison-Wesley Publishing Company.
Pfeiffer, P. (1978). *Concepts of Probability Theory* (Second Revised Edition). Mineola, NY: Dover Publications, Inc.

CHAPTER 5

Sports Prediction Models

5.1 INTRODUCTION

In this chapter, we provide an overview of six sports modeling techniques that will be applied to different sports including football, basketball, hockey, soccer, and baseball. Our goal is to provide readers with a step-by-step set of instructions to be able to formulate the model, estimate the parameters, and predict outcome variables.

It is also important to note that these models will also serve as the basis for predicting individual player performance, which could be used for fantasy sports competitions.

These models include:

1. Games Points
2. Team Statistics
3. Logistic Probability
4. Team Ratings
5. Logit Spreads
6. Logit Points

Readers interested in a more thorough understanding of the mathematics behind these models are referred to Chapter 2, Regression Models; Chapter 3, Probability Models; and Chapter 4, Advanced Math and Statistics.

These follow from the following four families of mathematical prediction models:

1. Linear Regression:

$$y = b_0 + b_1 \cdot x_1 + b_2 \cdot x_2 + \cdots + b_k \cdot x_k + e$$

2. Logistic Function:

$$Prob = \frac{1}{1 + \exp\left\{-\left(b_0 + b_h - b_a\right)\right\}}$$

Optimal Sports Math, Statistics, and Fantasy.
DOI: http://dx.doi.org/10.1016/B978-0-12-805163-4.00005-0

3. Power Function:

$$Prob = \frac{b_0 + b_h}{b_0 + b_h + b_a}$$

4. Logit Regression

$$b_0 + b_h - b_a = \log\left(\frac{p}{1 - p}\right)$$

5.2 GAME SCORES MODEL

Description

The game scores model is a linear regression model that predicts game outcomes based on the expected home team winning margin.

Model Form

The game scores model has the form:

$$Y = b_0 + b_1 \cdot HPS + b_2 \cdot HPA + b_3 \cdot APS + b_4 \cdot APA + \varepsilon$$

The variables of the model are:

Y = home team victory margin, i.e., home team score minus away team score. A positive value indicates the home team won the game by the indicated number of points, a negative value indicates the away team won the game by the indicated number of points, and a zero indicates that the game ended in a tie.

HPS = home team average points scored per game

HPA = home team average points allowed per game

APS = away team average points scored per game

APA = away team average points allowed per game

The parameters of the model, i.e., what we are trying to solve are:

b_0 = home field advantage value

b_1 = home team points scored parameter

b_2 = home team points allowed parameter

b_3 = away team points scored parameter

b_4 = away team points allowed parameter

ε = random noise

Solving the Model

The model parameters are estimated via ordinary least squares (OLS) regression analysis. This solution will provide us with the estimated parameters and regression statistics.

Analysts will need to evaluate the model using the regressions goodness of fit R^2, the t-stat showing the significance of the parameters and explanatory factors, the F-test showing the overall significance of the model structure, and the regression error SeY.

Estimating Home Team Winning Spread

The expected home team victory margin is calculated directly from the model parameters and input variables. For example, after solving the regression, we estimate the score as follows:

$$\hat{y} = \hat{b}_0 + \hat{b}_1 \cdot HPS_1 + \hat{b}_2 \cdot HPS + \hat{b}_3 \cdot HPA + \hat{b}_4 \cdot APS$$

If $\hat{y} > 0$ then the home team is expected to win by this value, if $\hat{y} < 0$ then the home team is expected to lose by this value (i.e., the away team is expected to win by this value), and if $\hat{y} = 0$ then the game is expected to end in a tie.

Estimating Probability
Calculating Probability That the Home Team Wins

The probability that the home team will win the game is determined from the expected home team winning value $\hat{Y}$ and the regression error SeY. Based on these two parameter values we can calculate the probability that the home team victory margin will be more than zero using the normal distribution. Analysts will need to ensure that the regression error term, however, does follow a normal distribution. And if not, probability calculations will need to be determined from the appropriate probability distribution function.

The probability p that the home team will win the game is computed as follows:

First, compute the normalized statistic:

$$z = \frac{0 - \hat{Y}}{SeY} = \frac{-\hat{Y}}{SeY}$$

Second, compute the probability that the home team wins the game:

$$p = 1 - F^{-1}(z)$$

Here, $F^{-1}(z)$ is the normal distribution inverse function for the standard normal variable z.

Alternatively, the probability that the home team will win the game can be computed directly using Excel or MATLAB functions as follows:
In Excel:

$$p = 1 - NormDist(0, \hat{Y}, SeY, True)$$

In MATLAB:

$$p = 1 - normcdf(0, \hat{Y}, SeY)$$

If $p > .50$ then the home team is predicted to win the game. If $p < .50$ then the away team is expected to win the game. If $p = .50$ then the game is expected to end in a tie.

Calculating Probability That the Home Team Wins by a Specified Score

We can also use the above techniques to compute the probability that the home team will win the game by more than a specified score S. This is determined from our regression results and from either Excel or MATLAB functions as follows:

In Excel:

$$Prob(Home\ Margin > S) = 1 - NormDist(S, \hat{Y}, SeY, True)$$

In MATLAB:

$$Prob(Home\ Margin > S) = 1 - normcdf\left(S, \hat{Y}, SeY\right)$$

5.3 TEAM STATISTICS MODEL

Description

The team statistics regression uses team performance statistics to predict game results. Generally speaking, these measurements could be either per-game averages (such as total yards per game) or per-event averages (such as yards per rush). Proper team performance statistics should also encompass both offensive and defensive ability. Through experimentation, readers can determine which set of team statistics provides the greatest predictive power.

Model Form

The team statistics linear regression model has the form:

$$Y = b_0 + b_1 \cdot HomeStat(1) + b_2 \cdot HomeStat(2) + b_3 \cdot HomeStat(3)$$
$$+ b_4 \cdot HomeStat(4) + b_5 \cdot AwayStat(1) + b_6 \cdot AwayStat(2)$$
$$+ b_7 \cdot AwayStat(3) + b_8 \cdot AwayStat(4) + \varepsilon$$

The variables of the model are:

Y = outcome value we are looking to predict (spread, home team points, away team points, and total points)

$HomeStat(1)$ = home team explanatory statistic 1
$HomeStat(2)$ = home team explanatory statistic 2
$HomeStat(3)$ = home team explanatory statistic 3
$HomeStat(4)$ = home team explanatory statistic 4
$AwayStat(1)$ = away team explanatory statistic 1
$AwayStat(2)$ = away team explanatory statistic 2
$AwayStat(3)$ = away team explanatory statistic 3
$AwayStat(4)$ = away team explanatory statistic 4
$b_0, b_1, b_2, b_3, b_4, b_5, b_6, b_7, b_8$ = model parameter, sensitivity to the corresponding explanatory team statistics variable
ε = random noise

Solving the Model

The model parameters are estimated via OLS regression analysis. This solution will provide us with the estimated parameters and regression statistics.

Analysts will need to evaluate the model using the regressions goodness of fit R^2, the t-stat showing the significance of the parameters and explanatory factors, the F-test showing the overall significance of the model structure, and the regression error SeY.

Estimating Home Team Winning Spread

The expected home team victory margin is calculated directly from the model parameters and input variables. For example, after solving the regression, we estimate the score as follows:

$$\hat{y} = \hat{b}_0 + \hat{b}_1 \cdot HomeStat(1) + \hat{b}_2 \cdot HomeStat(2) + \hat{b}_3 \cdot HomeStat(3)$$
$$+ \hat{b}_4 \cdot HomeStat(4) + \hat{b}_5 \cdot AwayStat(1) + \hat{b}_6 \cdot AwayStat(2)$$
$$+ \hat{b}_7 \cdot AwayStat(3) + \hat{b}_8 \cdot AwayStat(4)$$

If $\hat{y} > 0$ then the home team is expected to win by this value, if $\hat{y} < 0$ then the home team is expected to lose by this value (i.e., the away team is expected to win by this value), and if $\hat{y} = 0$ then the game is expected to end in a tie.

Estimating Probability

Calculating Probability That the Home Team Wins

The probability that the home team will win the game is determined from the expected home team winning value $\hat{Y}$ and the regression error SeY. Based on these two parameter values we can calculate the probability that the home team victory margin will be more than zero using the normal distribution. Analysts will need to ensure that the regression error term, however, does follow a normal distribution. And if not, probability calculations will need to be determined from the appropriate probability distribution function.

The probability p that the home team will win the game is computed as follows:

First, compute the normalized statistic:

$$z = \frac{0 - \hat{Y}}{SeY} = \frac{-\hat{Y}}{SeY}$$

Second, compute the probability that the home team wins the game is:

$$p = 1 - F^{-1}(z)$$

Here, $F^{-1}(z)$ is the normal distribution inverse function for the standard normal variable z.

Alternatively, the probability that the home team will win the game can be computed directly using Excel or MATLAB functions as follows:

In Excel:

$$p = 1 - NormDist(0, \hat{Y}, SeY, True)$$

In MATLAB:

$$p = 1 - normcdf(0, \hat{Y}, SeY)$$

If $p > .50$ then the home team is predicted to win the game. If $p < .50$ then the away team is expected to win the game. If $p = .50$ then the game is expected to end in a tie.

Calculating Probability That the Home Team Wins by a Specified Score

We can also use the above techniques to compute the probability that the home team will win the game by more than a specified score S. This is determined from our regression results and from either Excel or MATLAB functions as follows:

In Excel:

$$Prob(Home\ Margin > S) = 1 - NormDist(S, \hat{Y}, SeY, True)$$

In MATLAB:

$$Prob(Home\ Margin > S) = 1 - normcdf(S, \hat{Y}, SeY)$$

5.4 LOGISTIC PROBABILITY MODEL

Description

The logistic probability model is one of the more important probability models that can be applied to sports modeling. This technique allows us to infer a team strength rating based only on the observations of game outcomes. The most common set of outcomes is whether the team won, lost, or tied the game. The result of the game is based from the perspective of the home team but analysts can easily use the same approach based on the visiting team.

The traditional logistic probability model will define the outcome event as 1 if the home team won and 0 if the home team lost. While there are many approaches that statisticians can use to adjust for a tie, one suggested approach is to include the game in the dataset twice: once with the home team winning the game and the second with the away team winning the game.

Model Form

The logistic model is as follows:

$$G(x) = \frac{1}{1 + \exp\left\{-(b_0 + b_h - b_a)\right\}}$$

Here, b_0 denotes a home field advantage parameter, b_h denotes the home team rating parameter value, and b_a denotes the away team rating parameter value. Readers can easily verify that as the home team rating becomes much larger than the away team rating, the probability that the home team will win approaches 1, and as the home team rating becomes much smaller than the away team rating, the probability that the home team will win approaches zero.

The notation used to denote the outcome of the game for the logistic model is:

$$G(x) = \begin{cases} \dfrac{1}{1 + \exp\{-(b_0 + b_h - b_a)\}} & \text{if Home Team Wins} \\[4mm] \dfrac{1}{1 + \exp\{+(b_0 + b_h - b_a)\}} & \text{if Away Team Wins} \end{cases}$$

Notice that if the home team wins we use the logistic function $G(x)$ and if the home team loses we use $1 - G(x)$. The only difference in the reduced logistic function (after performing some algebra) is if the home team wins there is a negative sign in the exponent of $\exp\{\}$ in the denominator and if the home team loses there is a positive sign in the denominator of the exponential function.

Solving the Model

The solution to the logistic model is determined via maximum likelihood estimates (MLEs). Unfortunately, there is no simple adjustment or trick to enable us to solve this model via OLS regression techniques and the parameter values are most commonly determined via sophisticated optimization routines.

The MLE technique to solve the above is described in Chapter 3, Probability Models, and specified as:

$$Max\ L = \prod_{i=1}^{n} G_i(x)$$

where $G_i(x)$ is defined above for each game. This equation can also be rewritten in terms of a log transformation and using addition (as opposed to multiplication) as follows:

$$Max\ \text{Log}(L) = \sum_{i=1}^{n} \text{Log}(G_i(x))$$

Important Notes:

- The parameter value b_0 will be a constant for each game.
- The parameter values for the home team b_h and the away team b_a will be based on the actual teams playing in each game. For example,

if home team k plays away team j logistic equation is written as follows

$$P_i = \frac{1}{1 + \exp\left\{-\left(b_0 + b_k - b_j\right)\right\}}.$$

- The solution of the logistic regression results in a team strength-rating parameter for each team. This larger the team strength parameter, the better the team. These parameters can be used to directly determine the probability of winning a game and can also be used as input to a linear regression model such as the team rating model described above.

Estimating Home Team Winning Probability

After determining the team ratings from the MLEs we can compute the probability that the home team will win the game as follows:

$$Probability = \frac{1}{1 + \exp\left\{-\left(\hat{b}_0 + \hat{b}_h - \hat{b}_a\right)\right\}}$$

Estimating Home Team Winning Spread

Once this probability is known, the expected home team winning margin is computed from running a second regression analysis of spreads as a function of home team winning probability. To run this analysis, we need to compute the home team winning probability for each game.

Then, the estimated spread is determined from the following equation:

$$Home\ Team\ Spread = \hat{d}_0 + \hat{d}_1 \cdot Probability$$

where d_0, d_1 are determined from OLS analysis.

Important Notes:

- Analysts need to evaluate the regression error term to understand its distribution
- Based on actual data, we may find that a probability of $p = .50$ may have an expected home team winning spread that is different than zero. In theory, the expected home team spread corresponding to a probability level of $p = .50$ should be zero.
- If this occurs, analysts will need to make an adjustment to the model to correct for bias, or incorporate the appropriate error term distribution (see Chapter 4: Advanced Math and Statistics).

5.5 TEAM RATINGS MODEL

Description

The team ratings prediction model is a linear regression model that uses the team ratings determined from the logistic probability model as the explanatory variables to estimate the home team victory margin. This is one of the reasons why the logistic model is one of the more important sports models since its results can be used in different modeling applications.

Model Form

The team ratings regression has form:

$$Y = b_0 + b_1 \cdot Home\ Rating + b_2 \cdot Away\ Rating + \varepsilon$$

The variables of the model are:

Y = outcome value we are looking to predict (spread, home team points, away team points, and total points)

$Home\ Rating$ = home team strength rating (from logistic probability model)

$Away\ Rating$ = away team strength rating (from logistic probability model)

b_0 = home field advantage value

b_1 = home team rating parameter

b_2 = away team rating parameter

ε = random noise

Analysts can use this formulation to develop more complex linear and/or nonlinear models based on input from the logistic regression techniques.

Estimating Home Team Winning Spread

The expected home team victory margin is calculated directly from the model parameters and input variables. For example, after solving the regression, we estimate the score as follows:

$$\hat{Y} = \hat{b}_0 + \hat{b}_1 \cdot Home\ Rating + \hat{b}_2 \cdot Away\ Rating + \varepsilon$$

If $\hat{y} > 0$ then the home team is expected to win by this value, if $\hat{y} < 0$ then the home team is expected to lose by this value (i.e., the away team is expected to win by this value), and if $\hat{y} = 0$ then the game is expected to end in a tie.

Estimating Probability

Calculating Probability That the Home Team Wins

The probability that the home team will win the game is determined from the expected home team winning value $\hat{Y}$ and the regression error SeY. Based on these two parameter values we can calculate the probability that the home team victory margin will be more than zero using the normal distribution. Analysts will need to ensure that the regression error term, however, does follow a normal distribution. And if not, probability calculations will need to be determined from the appropriate probability distribution function.

The probability p that the home team will win the game is computed as follows:

First, compute the normalized statistic:

$$z = \frac{0 - \hat{Y}}{SeY} = \frac{-\hat{Y}}{SeY}$$

Second, compute the probability that the home team wins the game:

$$p = 1 - F^{-1}(z)$$

Here, $F^{-1}(z)$ is the normal distribution inverse function for the standard normal variable z.

Alternatively, the probability that the home team will win the game can be computed directly using Excel or MATLAB functions as follows:

In Excel:

$$p = 1 - NormDist(0, \hat{Y}, SeY, True)$$

In MATLAB:

$$p = 1 - normcdf(0, \hat{Y}, SeY)$$

If $p > .50$ then the home team is predicted to win the game. If $p < .50$ then the away team is expected to win the game. If $p = .50$ then the game is expected to end in a tie.

Calculating Probability That the Home Team Wins by a Specified Score

We can also use the above techniques to compute the probability that the home team will win the game by more than a specified score S. This is determined from our regression results and from either Excel or MATLAB functions as follows:

In Excel:

$$Prob(Home\ Margin > S) = 1 - NormDist(S, \hat{Y}, SeY, True)$$

In MATLAB:

$$Prob(Home\ Margin > S) = 1 - normcdf(S, \hat{Y}, SeY)$$

5.6 LOGIT SPREAD MODEL

Description

The logit spread model is a probability model that predicts the home team victory margin based on an inferred team rating metric and home team winning margin. The model transforms the home team victory margin to a probability value between zero and one and then the model can be solved via logit regression analysis.

The inferred team "ratings" are solved via a logit linear regression model. This technique was described in Chapter 3, Probability Models.

Model Form

The logit spread model has following form:

$$y^* = b_0 + b_h - b_a$$

where b_0 denotes a home field advantage parameter, b_h denotes the home team parameter value, and b_a denotes the away team parameter value.

The left-hand side of the equation y^* is the log ratio of the cumulative density function of victory margin and is computed as follows:

Step 1: $s_i =$ Home team victory margin in game i.

($s_i > 0$ indicates the home team won the game, $s_i < 0$ indicates the won team lost the game, and $s_i = 0$ indicates the game ended in a tie).

Step 2: Compute average home team victory margin, $\bar{s}$, across all games.

Step 3: Compute standard deviation of home team victory margin, σ_s, across all games.

Step 4: Compute the z-score of each spread, $z_i = (s_i - \bar{s})/\sigma_s$.

Step 5: Compute the cumulative probability corresponding to z_i as $F(z_i)$. This ensures the values of $F(z_i)$ will be between 0 and 1. This can be computed via Excel or MATLAB in the following manner:

Excel: $F(Z_i) = normsdist(z_i)$

MATLAB: $F(Z_i) = normcdf(z_i)$

Step 6: Compute the success ratio y^* as follows:

$$y^* = \frac{F(z_i)}{1 - F(Z_i)}$$

Step 7: Take the log transformation of y^* as follows:

$$y = \log(y^*) = \log\left(\frac{F(z_i)}{1 - F(Z_i)}\right)$$

Solving the Model

The team rating parameters b_k are then estimated via OLS regression analysis on the following equation:

$$y = \hat{b}_0 + \hat{b}_h - \hat{b}_a$$

Analysts need to perform an analysis of the regression results including evaluation of R^2, t-stats, F-value, and analysis of the error term.

Estimating Home Team Winning Spread

The home team expected winning spread is calculated as follows:

$$y = \hat{b}_0 + \hat{b}_k - \hat{b}_j$$

From the estimated y we determine the normalized z-score z for home team winning spread as follows:

$$z = F^{-1}\left(\frac{e^y}{1 + e^y}\right)$$

where F^{-1} represents the inverse of the normal distribution for the log-transformed home team winning spread.

This can be computed via Excel or MATLAB functions as follows:

Excel: $z = normsinv\left(\dfrac{e^y}{1 + e^y}\right)$

MATLAB: $z = norminv\left(\dfrac{e^y}{1 + e^y}\right)$

Finally, the expected home team expected points is calculated as follows:

$$\hat{s} = z \cdot \sigma_s + \bar{s}$$

where

$\bar{s}$ is the average home team winning spread

σ_s is the standard deviation of home team winning spread

Estimating Probability

The corresponding probability that the home team will win the game can be determined via a second regression analysis of actual spread as a function of estimated spread s. The model has the form:

$$Actual\ Spread = c_0 + c_1 \cdot \hat{s} + Error$$

Solution of this regression will provide model parameters $\hat{c}_0$ and $\hat{c}_1$ and regression error term $Error$.

After solving the second regression model and determining the regression error term $Error$, we can compute the probability that the home team will win the game.

It can be computed directly from the Excel or MATLAB functions as follows:

In Excel:

$$p = 1 - normdist(0, \hat{s}, Error, True)$$

In MATLAB:

$$p = 1 - normcdf(0, \hat{s}, Error)$$

If $p > .50$ then the home team is predicted to win the game. If $p < .50$ then the away team is expected to win the game. If $p = .50$ then the game is expected to end in a tie.

5.7 LOGIT POINTS MODEL

Description

The logit points model is a probability model that predicts the home team victory margin by taking the difference between expected home team points and expected away team points. The predicted points are determined based on inferred team "ratings" similar to the logit spread model discussed above.

Model Form

The logit points model has the following form:

$$h = c_0 + c_h - c_a$$
$$a = d_0 + d_h - d_a$$

where

h denotes is the transformed home team points, c_0 denotes a home field advantage parameter, c_h denotes the home team parameter value, and c_a denotes the away team parameter value.

a denotes the transformed away team points, d_0 denotes a home field advantage parameter, d_h denotes the home team parameter value, and d_a denotes the away team parameter value.

The left-hand side of the equation h, a represents the success ratio of the transformed points value and is computed as follows:

Step 1: h_i = Home team points in game i
a_i = Away team points in game i

Step 2: $\bar{h}$ = Average home team points across all games
$\bar{a}$ = Average away team points across all games

Step 3: σ_h = Standard deviation of home team points across all games
σ_a = Standard deviation of away team points across all games

Step 4: Compute the z-score of home and away points as follows:

$$\text{Home team: } z_{hi} = \left(h_i - \bar{h}\right)/\sigma_h$$
$$\text{Away team: } z_{ai} = (a_i - \bar{a})/\sigma_a$$

Step 5: Compute the standard normal cumulative probability for each values as follows:

Via Excel functions:

$$\text{Home team: } F_h(z_i) = normsdist(z_{Hi})$$
$$\text{Away team: } F_a(z_i) = normsdist(z_{Ai})$$

Via MATLAB functions:

$$\text{Home team: } F_h(z_i) = normcdf(z_{Hi})$$
$$\text{Away team: } F_a(z_i) = normcdf(z_{Ai})$$

Step 6: Compute the success ratios for home and away points:

$$\text{Home team points ratio: } h^* = \frac{F_h(z_{hi})}{1 - F_h(z_{hi})}$$

$$\text{Away team points ratio: } a^* = \frac{F_a(z_{ai})}{1 - F_a(z_{ai})}$$

Step 7: Take a log transformation of the success ratio:

$$\text{Home team points ratio: } h = \log(h^*) = \log\left(\frac{F_a(z_{ai})}{1 - F_a(z_{ai})}\right)$$

$$\text{Away team points ratio: } a = \log(a^*) = \log\left(\frac{F_a(z_{ai})}{1 - F_a(z_{ai})}\right)$$

5.8 ESTIMATING PARAMETERS

The team rating parameters b_k are then determined via OLS regression analysis following techniques in Chapter 2, Regression Models. These results will be shown for all sports in Chapter 6, Football—NFL; Chapter 7, Basketball—NBA; Chapter 8, Hockey—NHL; Chapter 9, Soccer—MLS; and Chapter 10, Baseball—MLB.

After we have the parameters for these models from OLS we can estimate the home and away points for the game via the following techniques:

Estimating Home Team Points

Estimating home team points is accomplished directly from the home team regression as follows. If team k is the home team and team j is the away team, the transformed home team points are:

$$h = \hat{c}_0 + \hat{c}_k - \hat{c}_j$$

From the estimated h we determine the normalized z-score z_h for home team points as follows:

$$z_h = F_h^{-1}\left(\frac{e^h}{1 + e^h}\right)$$

where F_h^{-1} represents the inverse of the normal distribution for the log-transformed home team points.

This can be computed via Excel or MATLAB functions as follows:

$$\text{Excel: } \quad z_h = normsinv\left(\frac{e^h}{1 + e^h}\right)$$

$$\text{MATLAB: } \quad z_h = norminv\left(\frac{e^h}{1 + e^h}\right)$$

Finally, the expected home team expected points is calculated as follows:

$$Est.\ Home\ Team\ Points = z_h \cdot \sigma_h + \overline{h}$$

where

$\overline{h}$ is the average home team winning points
σ_h is the standard deviation of home team points

Estimating Away Team Points

Estimating away team points is accomplished directly from the away team regression as follows. If team k is the home team and team j is the away team, the transformed away team points is:

$$a = \hat{c}_0 + \hat{c}_k - \hat{c}_j$$

From the estimated a we determine the normalized z-score z_a for away team points as follows:

$$z_a = F_a^{-1}\left(\frac{e^a}{1 + e^a}\right)$$

where F_a^{-1} represents the inverse of the normal distribution for the log-transformed away team points.

We can now determine z_a via Excel or MATLAB functions as follows:

Excel: $z_a = normsinv\left(\dfrac{e^a}{1 + e^a}\right)$

MATLAB: $z_a = norminv\left(\dfrac{e^a}{1 + e^a}\right)$

Finally, the estimated home team expected points is:

$$Est.\ Away\ Team\ Points = z_a \cdot \sigma_a + \overline{a}$$

where

$\overline{a}$ is the average away team winning points
σ_a is the standard deviation of away team points

Estimating Home Team Winning Spread

The estimated home team victory margin is computed directly from the home team points and away team points as follows:

$$Est.\ Spread = Est.\ Home\ Team\ Points - Est.\ Away\ Team\ Points$$

Estimating Home Team Winning Probability

The corresponding probability of winning is determined by performing a regression analysis of actual home team spread as a function of estimated spread (from above) The model has the form:

$$Actual\ Spread = w_0 + w_1 \cdot Est.\ Spread + Error$$

Solution of this regression will provide model parameters $\hat{w}_0$ and $\hat{w}_1$ and regression error term *Error*.

After solving the second regression model and determining the regression error term *Error*, we can compute the probability that the home team will win the game.

It can be computed directly from the Excel or MATLAB functions as follows:

In Excel:

$$p = 1 - NormDist(0, Est.\ Spread, Error, True)$$

In MATLAB:

$$p = 1 - normcdf(0, Est.\ Spread, Error)$$

If $p > .50$ then the home team is predicted to win the game. If $p < .50$ then the away team is expected to win the game. If $p = .50$ then the game is expected to end in a tie.

5.9 CONCLUSION

The chapter provided an overview of the six different sports prediction models that will be applied to NFL, NBA, MLS, NHL, and MLB in subsequent chapters. The models discussed in the chapter are based on linear regression, logistic probability, and logit regression models. For each of these, we provided a step-by-step process to estimate model parameters, predict home team winning spread, predict probability that the home team will win the game, and also the probability that the home team will win by a certain score or margin (such as a specified spread or margin).

CHAPTER 6

Football—NFL

In this chapter we apply our different sports modeling techniques to NFL football data for the 2015 season. Our goal is to provide readers with different techniques to predict winning team, estimated victory margin, and the probability of winning. These models are based on linear regression techniques described in Chapter 2, Regression Models, and on probability estimation methods described in Chapter 3, Probability Models. They include the games points, team statistics, team ratings, logistic probability, logit spreads, and logit points models. An overview of these approaches is also provided in Chapter 5, Sports Prediction Models.

We evaluated these models using in-sample data using different metrics: winning percentage, R^2 goodness of fit, and regression error. Our out-sample performance results are discussed at the end of the chapter. Overall, we found the best-performing models to be the probability-based models that infer explanatory factors based on the data.

Our results are as follows:

Fig. 6.1 depicts the winning percentage by model. In all cases except for one, our models had a better than 70% winning percentage. The team statistics model has a winning percentage of 68%. The team ratings model had the highest winning percentage of 75% followed by the logistic probability model. Each of these outperformed the Vegas line, which had a 64% winning percentage.

Fig. 6.2 illustrates the R^2 goodness of fit for each model. Each model had a relatively high goodness of fit for sports models with $R^2 >= 30\%$ for all but one of the models. The logit spread model and logit points model had the highest goodness of fit of $R^2 = 35\%$. The Vegas line had $R^2 = 16\%$.

Fig. 6.3 depicts the regression error surrounding the predicted victory margin. Here, the lower the regression error, the better the model fit. Our models had a regression error between 11 and 12 points, and the Vegas line had a regression error of 13.

Optimal Sports Math, Statistics, and Fantasy.
DOI: http://dx.doi.org/10.1016/B978-0-12-805163-4.00006-2

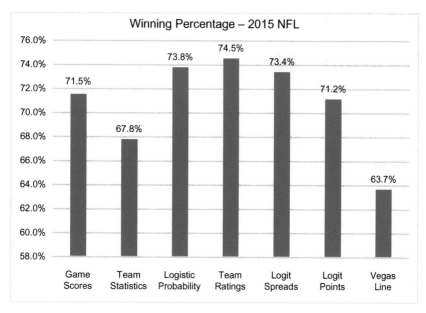

Figure 6.1 Winning Percentage: 2015 NFL.

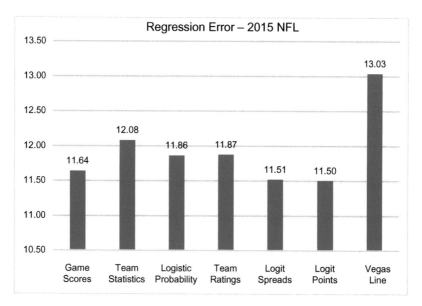

Figure 6.2 Regression Error: 2015 NFL.

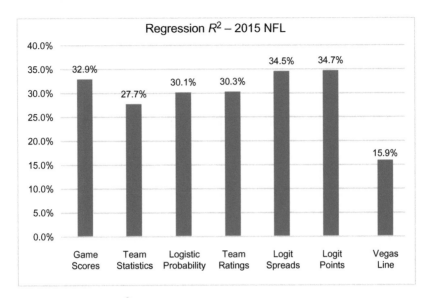

Figure 6.3 Regression R^2: 2015 NFL.

Table 6.1 provides the data used in these regression models to allow readers to test these models and experiment with different formulations. These models also include the rankings of the teams according to each of the six models described in the chapter.

In all cases, these models performed better than the Vegas line both in terms of how often the favorite won the game and how close the point spreads came to the actual margins of victory. It is, however, important to note that the Vegas line is a forward-looking model that moves in response to the wagers placed on the teams, and that its primary purpose is to ensure that the bookmakers make a profit regardless of the outcome by encouraging equal wagering on both teams. We are comparing "in-sample" data to Vegas results for comparison purposes. In the section titled Out-Sample Results (Section 6.8) we compare our look-forward results to the Vegas line and find that our results are online or better than these lines.

A description of each model is as follows:

6.1 GAME SCORES MODEL

The game scores regression model predicts game outcomes based on the average number of points scored and allowed by each team.

The model has the following form:

$$Y = b_0 + b_1 \cdot HPS + b_2 \cdot HPA + b_3 \cdot APS + b_4 \cdot APA + \varepsilon$$

Table 6.1 Regression Results and Rankings

Team		Team Statistics							Ratings				Rankings				
	W – L	PF/G	PA/G	OY/R	OY/PA	DY/R	DY/PA	Logistic Rating	Logit Spread	Logit Home Points	Logit Away Points	Game Scores	Team Statistics	Logistic Ratings	Logit Spread	Logit Points
Arizona Cardinals	14 – 4	29.4	21.2	4.14	8.07	4.04	6.64	3.368	2.826	2.233	0.346	8	1	3	4	2
Atlanta Falcons	8 – 8	21.2	21.6	3.81	7.06	4.04	6.92	1.160	1.035	1.441	2.101	15	10	20	22	21
Baltimore Ravens	5 – 11	20.5	25.1	3.86	6.32	3.97	6.86	0.735	1.320	1.344	1.678	22	21	24	17	18
Buffalo Bills	8 – 8	23.7	22.4	4.78	7.19	4.39	6.60	1.258	1.576	1.430	1.428	13	12	18	13	13
Carolina Panthers	17 – 2	36.9	19.4	4.27	7.34	3.91	5.90	3.937	2.792	2.621	0.862	1	4	1	5	5
Chicago Bears	6 – 10	20.9	24.8	3.96	7.00	4.47	7.00	1.468	1.405	1.564	1.858	21	13	15	16	16
Cincinnati Bengals	12 – 5	27.2	17.5	3.86	7.57	4.41	6.18	2.777	2.850	2.219	0.397	5	2	4	3	4
Cleveland Browns	3 – 13	17.4	27.0	4.02	6.21	4.49	7.85	0.059	0.786	1.102	2.191	32	32	31	24	25
Dallas Cowboys	4 – 12	17.2	23.4	4.63	6.56	4.20	7.19	0.226	0.714	1.634	2.936	29	27	29	26	29
Denver Broncos	15 – 4	26.4	17.9	4.02	6.46	3.36	5.81	3.447	2.449	2.070	0.746	6	8	2	8	8
Detroit Lions	7 – 9	22.4	25.0	3.77	6.66	4.22	7.18	1.734	1.519	1.807	1.873	19	20	11	14	14
Green Bay Packers	11 – 7	26.4	20.4	4.34	6.09	4.44	6.77	2.672	2.265	1.893	0.811	9	28	7	10	9
Houston Texans	9 – 8	21.2	20.2	3.71	6.08	4.09	6.18	1.424	1.310	0.804	1.123	12	14	16	18	17
Indianapolis Colts	8 – 8	20.8	25.5	3.63	5.98	4.32	7.03	1.239	0.663	0.475	1.689	24	26	19	29	27
Jacksonville Jaguars	5 – 11	23.5	28.0	4.16	6.77	3.68	7.13	0.172	0.641	0.541	1.856	27	19	30	30	30
Kansas City Chiefs	12 – 6	28.9	17.1	4.56	6.98	4.14	5.90	2.674	2.857	2.211	0.377	2	7	6	2	3
Miami Dolphins	6 – 10	19.4	24.3	4.35	6.48	4.01	7.38	0.521	0.712	0.943	2.112	23	29	28	27	26
Minnesota Vikings	11 – 6	23.4	18.4	4.51	6.43	4.20	6.65	2.654	2.278	2.243	1.211	10	23	8	9	10
New England Patriots	13 – 5	31.4	20.1	3.66	7.00	3.89	6.46	2.518	2.507	2.271	0.912	4	6	10	7	7
New Orleans Saints	7 – 9	25.4	29.8	3.76	7.45	4.91	8.35	0.955	0.694	1.779	3.075	28	24	23	28	28
New York Giants	6 – 10	26.3	27.6	3.99	6.98	4.37	7.50	0.704	1.075	1.292	2.001	16	22	25	21	22
New York Jets	10 – 6	24.2	19.6	4.17	6.72	3.58	6.26	1.585	1.734	1.804	1.555	11	9	13	11	11
Oakland Raiders	7 – 9	22.4	24.9	3.94	6.40	4.13	6.46	1.520	1.591	1.652	1.611	18	15	14	12	12
Philadelphia Eagles	7 – 9	23.6	26.9	3.93	6.56	4.50	6.75	1.002	0.931	1.111	2.004	20	17	22	23	23
Pittsburgh Steelers	11 – 7	28.6	19.9	4.53	7.82	3.78	6.81	2.539	2.665	2.529	0.999	7	5	9	6	6
San Diego Chargers	4 – 12	20.0	24.9	3.46	6.88	4.81	7.42	0.523	1.292	1.410	1.855	25	18	27	19	20
San Francisco 49ers	5 – 11	14.9	24.2	3.96	6.30	4.01	7.61	1.060	0.776	0.374	1.422	31	31	21	25	24
Seattle Seahawks	11 – 7	28.6	17.6	4.52	7.63	3.49	6.18	2.711	2.931	2.720	0.785	3	3	5	1	1
St. Louis Rams	7 – 9	17.5	20.6	4.56	5.93	4.02	6.81	1.631	1.476	1.527	1.646	17	30	12	15	15
Tampa Bay Buccaneers	6 – 10	21.4	26.1	4.76	7.20	3.45	7.10	0.605	0.598	1.037	2.379	26	16	26	31	31
Tennessee Titans	3 – 13	18.7	26.4	4.00	6.36	3.89	7.31	-0.448	0.257	0.604	2.450	30	25	32	32	32
Washington Redskins	9 – 8	25.4	24.4	3.69	7.36	4.80	7.21	1.360	1.252	1.282	1.700	14	11	17	20	19
Home Field Advantage								0.212	0.223	0.034	0.012					

In this representation, the dependent variable Y denotes the home team's margin of victory (or defeat). A positive value indicates the home team won by the stated number of points and a negative value indicates the home team lost by the stated number of points.

The variables of this model are:

Y = home team victory margin
HPS = home team average points scored per game
HPA = home team average points allowed per game
APS = away team average points scored per game
APA = away team average points allowed per game
b_0, b_1, b_2, b_3, b_4 = model parameters, represents the sensitivities to the corresponding model factor.

The model parameters, i.e., betas, are determined from a linear regression analysis as described in Chapter 2, Regression Models.

The probability that the home team will be victorious is computed from the Excel normal distribution function as follows:

$$p = 1 - NormDist(0, Y, SeY, True)$$

Winning probability can also be computed from a statistical package such as via MATLAB's normal cumulative distribution function as follows:

$$p = 1 - normcdf(0, Y, SeY)$$

Here, Y and SeY are the expected victory margin and regression error term respectively, and zero indicates the reference point used for the calculation (i.e., the probability that the winning margin will be greater than zero). This notation can be written more generally as:

Regression Results

The best fit regression equation for predicting the home team victory margin based on average points scored and average points allowed is below. Victory margin is positive when the home team is favored and negative when the visiting team is favored.

This model equation is:

$$Victory\ Margin = -1.28 + 0.867 \cdot HPS - 0.823 \cdot HPA - 0.609 \cdot APS$$
$$+ 0.999 \cdot APA \pm 11.72$$

The game scores model returned a relatively high R^2 value of $R^2 = 33\%$. The t-stats were -3.47 and 4.20 for visiting teams' points scored and points allowed, and 4.99 and -3.48 for home teams, and an F-value of 32.08. The regression error was 11.72.

The signs of the input variables are intuitive. The sign of the parameter is expected to be positive for home points scored (HPS) and away points allowed (APA). In both of these cases, the home team will score more points if these input values are higher. Similarly, the signs of the parameters for home points allowed (HPA) and away team points scored (APS) are expected to be negative. The home team is expected to win by fewer points or possibly lose if the HPA and/or APS increase in value.

Table 6.2 shows the results of the regression model and Fig. 6.4 shows a graph of actual victory margin (spreads) compared to estimated victory margin (spreads) from our model.

Table 6.2 Game Scores Model: 2015 NFL		
Statistic	**Value**	**t-Stat**
b_0	−8.519	−0.8
b_1 (Home points scored)	0.867	5.0
b_2 (Home points allowed)	−0.823	−3.5
b_3 (Visitor points scored)	−0.609	−3.5
b_4 (Visitor points allowed)	0.999	4.2
R^2	32.88%	
F-value	32.082	
Standard Error	11.724	

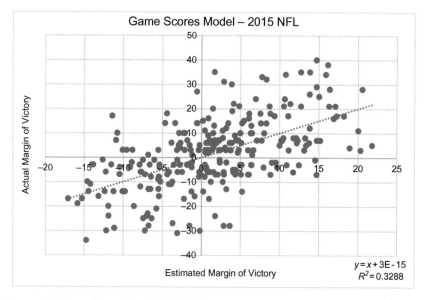

Figure 6.4 Game Scores Model: 2015 NFL.

Performance

The game scores model correctly identified the winner in 190 of 267 games, a rate of 71.2%. When the model gave the favorite at least a 60% win probability, the favorite won 78.5% of the time, and that figure jumped to 84.2% when the model predicted one team or the other as having a probability of winning of 70% or higher. Above 85%, the model was wrong only 3 times out of 46: a three-point loss by Baltimore to Pittsburgh in Week 4 in which the model had the Ravens as a 6-to-1 favorite, a two-point loss by Kansas City to Chicago in Week 5 in which the Bears were projected to be a 15½-point underdog, and a 35−28 defeat of the New England Patriots in Week 13 at the hands of the Philadelphia Eagles, who made the most of their 10.5% probability of winning.

For the 2015 NFL postseason, the model correctly predicted the winner in 9 of 11 games, missing only the Bengals' loss to the Steelers in the first AFC Wild Card Game and the Broncos' upset of the Panthers in Super Bowl 50, in which it had considered Carolina an 8-to-5 favorite.

The Vegas model predicted the winner 170 times out of 267 games for a winning percentage of 63.7%. The game scores regression predicted the winner 190 times out of 267 games for a winning percentage of 71.2%. Thus the game scores regression outperformed the Vegas line and won 21 more games.

The game scores regression won 66.3% of simulated spread bets against the Vegas line over the course of the 2015 NFL season, a record of 171−87−9. If the Vegas line was the Patriots by 9, the model bet on the Patriots when it predicted they would win by more than 9 points, and against the Patriots when it predicted they would win by fewer than 9 points or lose outright.

In the 2015 NFL postseason, the model won 9 of 11 times against the Vegas line, including both conference championship games. The only missed projections were the division game between Arizona and Green Bay and the Super Bowl, in which Vegas had the Panthers as 5-point favorite whereas the model projected a 6.7-point win by Carolina.

Rankings

The top 10 in the game scores model's rankings all made the playoffs, as well as 11 of the top 12. The one playoff team not in the top 12 was the Washington Redskins, who ranked No. 14. The highest-ranked nonplayoff team was the New York Jets (10−6), who finished behind the 12−4 New England Patriots in the AFC East and lost the tiebreaker with the Pittsburgh Steelers for the AFC's second wild card slot. The NFC champion Panthers were the No. 1 ranked team while the Super Bowl champion Broncos came in at No. 6, between the Cincinnati Bengals and the Steelers.

Example

We compare a game scenario between the Dallas Cowboys (4−12) at home versus the Oakland Raiders (7−9).

The 2015 Cowboys scored 17.2 points per game while allowing 23.4. The Raiders scored 22.4 points per game while their defense gave up 24.9 points per game. Given those averages and our regression results, the game scores model favors the Raiders in this game by 1.59 points:

$$Victory\ Margin = -8.52 + 0.867(17.2) - 0.823(23.4) - 0.609(22.4)$$
$$+ 0.999(24.9) = -1.59$$

The negative victory margin in this case indicates that the visiting team (Raiders) is the favored team by 1.59.

The corresponding probability that the host Cowboys would win is 45% and is calculated as follows:

$$p = 1 - NormCDF(0, -1.59, 11.72) = 44.62\%$$

6.2 TEAM STATISTICS MODEL

The team statistics regression uses team performance statistics (i.e., team data) to predict game results.

We will demonstrate the regression approach using four team statistics per team: yards gained per rushing attempt and per pass attempt by the offense, and yards allowed per rush and per pass attempt by the defense.

The team statistics linear regression model has form:

$$Y = b_0 + b_1 \cdot HTOYR + b_2 \cdot HTOYPA + b_3 \cdot HTDYR + b_4 \cdot HTDYP$$
$$+ b_5 \cdot ATOYR + b_6 \cdot ATOYPA + b_7 \cdot HTDYR + b_8 \cdot ATDYPA + \varepsilon$$

Table 6.3 Team Statistics Model: 2015 NFL		
Statistic	**Value**	**t-Stat**
b_0	21.795	0.0
b_1 (Home off yards/rush)	-1.635	0.8
b_2 (Home off yards/pass attempt)	4.349	-0.7
b_3 (Home def yards/rush)	0.258	3.1
b_4 (Home def yards/pass attempt)	-6.931	0.1
b_5 (Away off yards/rush)	-2.555	-4.6
b_6 (Away off yards/pass attempt)	-4.246	-1.1
b_7 (Away def yards/rush)	-4.886	-3.1
b_8 (Away def yards/pass attempt)	9.179	-2.0
R^2	27.67%	
F-value	12.336	
Standard Error	11.724	

The variables of the model are:

Y = outcome value we are looking to predict (spread, home team points, away team points, and total points)

$HTOYR$ = home team average yards per rush

$HTOYPA$ = home team average yards per pass attempt

$HTDYR$ = home team average yards per rush allowed

$HTDYPA$ = home team average yards per pass attempt allowed

$ATOYR$ = away team average yards per rush

$ATOYPA$ = away team average yards per pass attempt

$ATDYR$ = away team average yards per rush allowed

$ATDYPA$ = away team average yards per pass attempt allowed

$b_0, b_1, b_2, b_3, b_4, b_5, b_6, b_7, b_8$ = model parameter, sensitivity to the variable

ε = model error

The betas of this model are determined from a linear regression analysis as described in Chapter 2, Regression Models. The results of this model are shown in Table 6.3.

Regression Results

The result of the regression model for predicting home team victory margin from our team statistics is:

$$Est.\ Victory\ Margin = 21.795 - 1.635(HTOYR) + 4.349(HTOYPA)$$
$$+ 0.258(HTDYR) - 6.931(HTDYPA)$$
$$- 2.555(ATOYR) - 4.246(ATOYPA)$$
$$- 4.886(ATDYR) + 9.179(ATDYPA) \pm 12.26$$

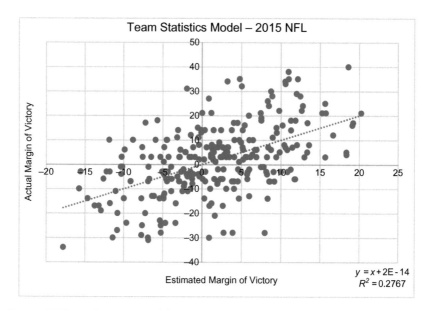

Figure 6.5 Team Statistics Model: 2015 NFL.

These regression results had a goodness of fit of $R^2 = 27.7\%$. The t-statistics were significant for passing but not significant for rushing. Furthermore, the passing relationship is positive while the rushing variable is negative. This does not imply that better rushing teams result in less performance. Rather, it implies that the teams that rush more often lose out on the better yards per pass completion. These results indicate that passing is more valuable than rushing to a team.

The high F-value of $F = 12.34$ is an indication that these variables are all not simultaneously zero. These regression statistics lend some support that analysts may be able to fine tune and improve the model using addition or different data variables as the input factors.

The regression standard deviation was 12.26. This is better than the Vegas line at 13.03. Our predicted results were within 1 point of the actual margin of victory 16 times (5.99%), 3 points 56 times (20.97%), and 7 points 126 times (47.19%).

These regression results are shown in Table 6.3 and the graph of actual spread compared to predicted spread is shown in Fig. 6.5.

Performance

The team statistics model predicted the winner 181 times in 267 games (67.8%). When the win probability was at least 60% the model was correct 74.59% of the time, and when the win probability was 80% or higher the model was correct 90.74% of the time. The model was a perfect 23 for 23 when the win probability was 85% or higher. For Super Bowl 50, this model predicted the Panthers as a 3-to-2 favorite.

The Vegas model predicted the winner 170 times out of 267 games for a winning percentage of 63.7%, so the team statistics model outperformed the Vegas line by 11 games. Standard deviations for the point spreads relative to the home team were 6.20 points for the Vegas line and 7.47 points for the team statistics model.

The team statistics model beat the Vegas line in 63.95% of the games, a record of 165−93−9. It also won simulated spread bets for 7 of the 11 playoff games, including both conference championships and the Super Bowl. Both the model and Vegas favored the Panthers, but the model would have advised taking the Broncos as its predicted victory margin of 4.6 points would not have covered the official Vegas spread of 5 points.

Rankings

The top eight were all playoff teams, with the next playoff team coming in at No. 11 (Washington). The Chicago Bears, who finished last in the NFC North at 6−10, were ranked No. 13, one spot ahead of the AFC South champion Houston Texans. The two playoff representatives from the NFC North finished far down the rankings, with the champion Minnesota Vikings ranked No. 23 and the wild card Green Bay Packers at No. 28.

The Cardinals were ranked No. 1 by this model, helped in large part by their 8.07 yards per offensive pass attempt, which was at least half a yard better than all other teams except the Steelers (7.82) and Seahawks (7.63). The Panthers finished in the No. 4 slot just a fraction behind Seattle, with the Broncos coming in at No. 8. These ranking results are shown in Table 6.1.

Example

The Cowboys and Raiders had nearly identical yards gained per pass attempt (Oakland 6.40, Dallas 6.56) and yards allowed per rush (Oakland 4.13, Dallas 4.20), while the Cowboys had an advantage on rushing offense (4.63 to 3.94) as did the Raiders on pass defense (6.46 to 7.19). But as the model's beta value for pass defense (-6.93) was much more significant than that for rushing offense (-1.64). Predictions from this model find Oakland as a 4-point favorite:

$$Est\ Margin\ of\ Victory = 21.795 - 1.635(4.63) + 4.349(6.56) + 0.258(4.20)$$
$$- 6.931(7.19) - 2.555(3.94) - 4.246(6.40)$$
$$- 4.886(4.13) - 9.179(6.46) \pm SE = -4.09 \pm 12.26$$

The estimated victory margin of -4.09 points is equivalent to a win probability of 36.9% for the Cowboys:

$$p = 1 - NormCDF(0, -4.09, 12.26) = 36.94\%$$

6.3 LOGISTIC PROBABILITY MODEL

The logistic probability model infers a team strength rating based only on game outcomes such as whether the team won, lost, or tied the game. The result of the game is determined from the perspective of the home team, but analysts can use the same approach from the perspective of the visiting team.

The logistic model is as follows:

$$y^* = \frac{1}{1 + \exp\{-(b_0 + b_h - b_a)\}}$$

Here, b_0 denotes a home field advantage parameter, b_h denotes the home team rating parameter value, and b_a denotes the away team rating parameter value. The value y^* denotes the probability that the home team will win the game. Team ratings for the logistic probability are determined via maximum likelihood estimates and are shown Table 6.1.

Estimating Spread

The estimated spread (i.e., home team victory margin) is determined via a second analysis where we regress the actual home team spread on the estimated probability y^* (as the input variable). This regression has form:

$$Actual\ Spread = a_0 + a_1 \cdot y^*$$

This model provides a relationship between the logistic home team winning probability and the home team winning percentage. It is important to note here that analysts may need to incorporate an adjustment to the spread calculation if the data results are skewed (see Chapter 3: Probability Models).

The solution to this model is (Table 6.4):

Table 6.4 Logistic Probability Regression: 2015 NFL		
Statistic	Value	t-Stat
a_0	−14.467	−8.7
a_1	29.585	10.7
b_0 (Home field advantage)	0.212	
R^2	30.25%	
F-value	114.947	
Standard Error	11.883	

After computing the home team winning probability y^*, the expected winning spread is estimated from the following equation using the regression results:

$$Estimated\ Spread = -14.48 + 29.59 \cdot y^*$$

A graph illustrating the estimated spread from the probability estimates is shown in Fig. 6.6.

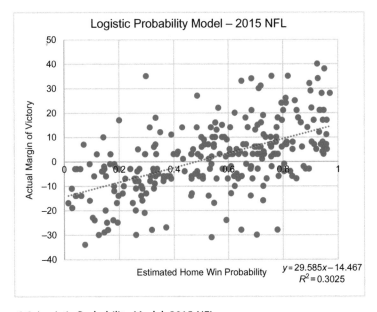

Figure 6.6 Logistic Probability Model: 2015 NFL.

Performance

The logistic probability model predicted the winning team in 197 of 267 games for a winning percentage of 73.8%. This is an improvement over the Vegas line (which predicted 170 games out of 267 correctly) by 27 games.

The logistic probability model also proved to be a very good predictor of the victory margin with an $R^2 = 30.29\%$ and standard error of 11.87. The standard error of the Vegas line was 13.05.

The model's favorites won 85.5% of the time when their win probability was 75% or higher. Above 90% it was wrong only once, in the case of the Panthers' Week 16 loss to Atlanta; the model returned a 92.9% win probability for Carolina. The median win probability for the favorite was 72.8%; the home team was the favorite 59.8% of the time.

In Super Bowl 50, this model predicted Carolina as a favorite over Denver with a win probability of 62%. The estimated winning spread for Carolina was 3.9 points. Had it been a home game for Carolina rather than a neutral-site game, their win probability would rise to 66.9% with a spread of 5.32 points.

Rankings

All of the teams in the model's top 10 rankings were among the 12 that qualified for the playoffs. The only two playoff teams not in the model's top 12 were the Houston Texans and Washington Redskins, who were ranked No. 16 and No. 17 respectively. These teams finished at 9−7 but were each in a division where they were the only team to finish above .500, thus qualifying for the NFL playoffs. The top teams in the model's rankings who were left out of the postseason were the Detroit Lions and the St. Louis Rams. Both finished with a records of 7−9, followed by the New York Jets (10−6).

The 2015 Lions' opponents averaged a ranking of 12.19; by that measure Detroit had the fourth-toughest schedule in the NFL. Half of their schedule came against teams that finished in the top eight—the Broncos, Cardinals, Chiefs, Seahawks, and two games each against the Vikings and Packers—against whom they went 1−7, their sole bright spot having been an 18−16 win over Green Bay in Week 10.

The Rams had a similar schedule. Their average opponent was ranked No. 13.125, seventh-toughest, and half their schedule came against teams that finished in the top nine. St. Louis did fare slightly better in those games, having gone 3−5, including a two-point win over the No. 3 Cardinals in Week 4 and a sweep of the season series with the No. 5 Seahawks.

The Jets were one of 6 AFC teams with 10 wins; unfortunately for Gang Green, each conference has only five playoff berths, and the AFC's wild cards were the Kansas City Chiefs (11−5) and Pittsburgh Steelers (10−6). The Jets had the weakest schedule in the NFL with an average opponent ranking of 21.69 and only two games against the top 10, both times the No. 10 Patriots, with each squad claiming one victory. Nine of the Jets' 10 wins came against teams in the bottom half.

The No. 1 Panthers actually had the second-easiest schedule, with 13 of 16 opponents in the bottom half of the rankings and 8 in the bottom 10, but 15−1 is still 15−1. Carolina's opponents in Super Bowl 50, the champion Denver Broncos, ranked No. 2.

Example

The Cowboys' logistic rating was the fourth-lowest in the NFL at 0.2256. Oakland's rating of 1.5202, meanwhile, ranked 14th out of the 32 teams. The logistic probability model gives the Cowboys only a 25.3% probability of beating the Raiders at home:

$$y^* = \frac{1}{1 + e^{-(0.2256-1.5202+0.2232)}} = \frac{1}{1 + e^{1.082}} = 0.2531$$

The estimated winning margin is computed using its value ($y^* = 0.2531$) and the regression parameters from above. This calculation results in a nearly 7-point victory by the visiting Raiders.

$$\text{Estimated Spread} = -14.48 + 29.59 \cdot 0.2531 = -6.98$$

6.4 TEAM RATINGS MODEL

The team ratings prediction model is a linear regression model that uses the team ratings determined from the logistic probability model as the explanatory variables to estimate home team victory margin. This is one of the reasons why the logistic model is one of the more important sports models since its results can be used in different modeling applications.

The team ratings regression has form:

$$Y = b_0 + b_1 \cdot Home\ Rating + b_2 \cdot Away\ Rating + \epsilon$$

The variables of the model are:

Home Rating = home team strength rating (from logistic probability model)

Away Rating = away team strength rating (from logistic probability model)

b_0 = home field advantage value

b_1 = home team rating parameter

b_2 = away team rating parameter

Y = home team's victory margin (positive indicates home team is favored and negative value indicates away team is favored)

The probability of winning is determined from:

$$Probability = 1 - NormCDF(0, Y, SE)$$

The betas of this model, b_1 and b_2, are determined from a linear regression analysis as described in Chapter 2, Regression Models.

Regression Results

The best fit regression equation to predict home team victory margin from team strength ratings is:

$$Estimated\ Victory\ Margin = 0.9925 + 5.5642 \cdot Home\ Rating$$
$$- 5.2037 \cdot Away\ Rating$$

The regression had an R^2 of 30% (considered high for sports prediction models) and significant t-stats for each of the team rating parameters. The standard error of this model is 11.92, indicating that 70% of the time the actual spread will be within 11.92 points of our predicted spread.

The regression results are shown in Table 6.5 and a graph showing the actual victory margin as a function of estimates victory margin is shown in Fig. 6.7.

Table 6.5 Team Ratings Model: 2015 NFL		
Statistic	**Value**	**t-Stat**
b_0	0.992	0.6
b_1 (Home team rating)	5.564	8.3
b_2 (Away team rating)	−5.204	−7.7
R^2	30.11%	
F-value	56.868	
Standard Error	11.918	

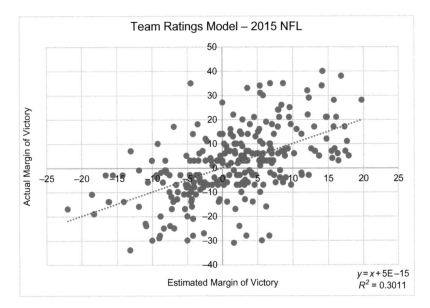

Figure 6.7 Team Ratings Model: 2015 NFL.

Performance

For the 2015 NFL season and postseason, the team ratings model correctly returned the winner in 199 of 267 games (74.53%). This success rate increased as the probability confidence increased; when the probability of winning is at least 65%, the model correctly predicted the winning team 80% of the time. At a predicted winning probability rate of 90%, the model was correct in 35 out of 36 games. Unsurprisingly, that one incorrect prediction was the Carolina Panthers' 20−13 loss to the Atlanta Falcons in Week 16, which ended the Panthers' bid for a perfect season and dropped their record to 14−1.

The Vegas model predicted the winner 170 times out of 267 games for a winning percentage of 63.7%, so the team ratings model outperformed the Vegas line by 29 games. The team ratings model beat the Vegas line in 67.4% of simulated spread bets, winning 174 wagers while losing 84. The team ratings model was also 10 for 11 in bets on playoff games including both conference championships and the Super Bowl. Of the six models, the team ratings model favored Carolina over Denver by 2.64 points in the Super Bowl.

Example

Given the Cowboys' rating of 0.2256 and the visiting Raiders' rating of 1.5202, the team ratings model would predict a victory by Oakland by 5.66 points.

$$\textit{Est. Victory Margin} = 0.9925 + 5.5642(0.2256) - 5.2037(1.5202) = -5.66$$

The estimated winning probability is calculated using the estimated victory margin and regression error terms. This results in a 31.7% probability for Dallas and thus 68.3% for Oakland:

$$\textit{Probability} = 1 - \textit{NormCDF}(0, -5.66, 11.92) = 31.7\%$$

6.5 LOGIT SPREAD MODEL

The logit spread model is a probability model that predicts the home team victory margin based on an inferred team rating metric. The model transforms the home team victory margin to a probability value between 0 and 1 via the cumulative distribution function and then estimates model parameters via logit regression analysis.

The logit spread model has following form:

$$y = b_0 + b_h - b_a$$

where b_0 denotes a home field advantage parameter, b_h denotes the home team parameter value, and b_a denotes the away team parameter value.

The left-hand side of the equation y is the log ratio of the cumulative density function of victory margin (see chapter: Sports Prediction Models for calculation process).

In this formulation, the parameters values b_0, b_h, b_a are then determined via ordinary least squares regression analysis. The results of this analysis are shown in Table 6.1.

Estimating Spreads

Estimating the home team winning margin is accomplished as follows.

If team k is the home team and team j is the away team, we compute y using the logit parameters:

$$y = b_0 + b_k - b_j$$

Compute y^* from y via the following adjustment:

$$y^* = \frac{e^y}{1 + e^y}$$

Compute z as follows:

$$z = \textit{norminv}(y^*)$$

And finally, the estimated home team spread:

$$Estimated\ Spread = \bar{s} + z \cdot \sigma_s$$

where

$\bar{s}$ = average home team winning margin (spread) across all games

σ_s = standard deviation of winning margin across all games

Estimating Probability

The corresponding probability of the home team winning is determined by performing a regression analysis of actual spread as a function of estimated spread to determine a second set of model parameters. This model has form:

$$Actual\ Spread = a_0 + a_1 \cdot Estimated\ Spread$$

To run this regression, we need to compute the estimated spread for all games using the logit spread parameters from above (see Table 6.1). The solution to this model is (Table 6.6):

Table 6.6 Logit Spread Model: 2015 NFL	
Statistic	**Value**
$\bar{s}$ (Average home victory margin)	1.599
σ_s (Home victory margin standard deviation)	14.202
b_0 (Home field advantage)	0.223
R^2	34.50%
F-value	139.553
Standard Error	11.516

A graphical illustration of the relationship between actual spreads and estimated spreads is shown in Fig. 6.8.

Performance

The logit spread model predicted the winner in 197 of 267 games (73.8%). It was 76% accurate when the favorite's probability of winning was over 60%, and 81% correct when the probability was 75% or higher. With a win probability of 97.1% and a spread of 21.8 points, the biggest favorite was the New England Patriots when they hosted the Tennessee Titans in Week 15, winning 33—16. The most even matchup involved the Cincinnati Bengals at the Denver Broncos in Week 16, for which the model calculated a win probability of 50.08% for Denver. The logit spread model's margin of victory regression had the highest R^2 with $R^2 = 34.604\%$, and standard regression error of 11.5.

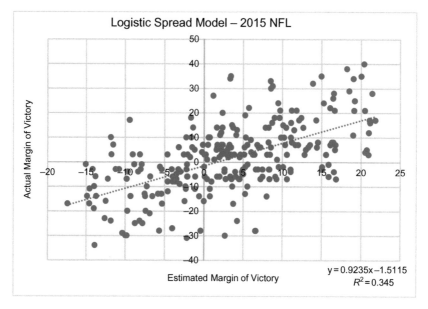

Figure 6.8 Logit Spread Model: 2015 NFL.

This model correctly predicted the winners of 9 of the 11 playoff games, missing only the AFC Wild Card Game between Cincinnati and Pittsburgh and the Super Bowl. In Super Bowl 50, the logit spread model gave the Panthers a 55% chance of beating the Broncos, making them a 1.45-point favorite, which was well under the Vegas line of Carolina by 5.

Rankings

The 12 playoff teams consisted of the top 10 in the logit spread rankings, followed by the Texans and Redskins, who were ranked 16 and 17 respectively. The top two in the rankings were the two conference champions, the Panthers and the Broncos.

Example

The Cowboys' logit spread parameter was 0.7139 (26th of the 32 teams), while the Raiders yielded a rating of 1.5912, good for 12th in the NFL. The home field advantage factor (b_0) for the NFL in 2015 was 0.2232. The average winning margin was $+1.6$ and the standard deviation of winning margin was 14.2.

The logit spread model predicts Oakland to win by 4.17 with a winning probability of 35.83%.

6.6 LOGIT POINTS MODEL

The logit points model is a probability model that predicts the home team victory margin by taking the difference between home team predicted points and away team predicted points. The predicted points are determined based on inferred team "ratings" similar to the logit spread model discussed above.

The logit points model has following form:

$$h = c_0 + c_h - c_a$$
$$a = d_0 + d_h - d_a$$

where

> h is the transformed home team points, c_0 denotes a home field advantage, c_h denotes the home team rating, and c_a denotes the away team rating corresponding to home team points.
>
> a is the transformed away team points, d_0 denotes a home field advantage, d_h denotes the home team rating, d_a denotes the away team rating corresponding to away team points.

The left-hand side of the equation h and a is the log ratio of the cumulative density function of home team points and away team points respectively (see chapter: Sports Prediction Models for a description).

Estimating Home and Away Team Points

Estimating the home team points is accomplished directly from the home points team ratings. These rating parameters are shown in Table 6.1. If team k is the home team and team j is the away team, the transformed home team points is:

$$h = c_0 + c_h - c_a$$

$$h^* = \frac{e^h}{1 + e^h}$$

Then

$$z = norminv(h^*)$$

And finally, home team estimated points are calculated as:

$$Home\ Team\ Points = \overline{h} + z \cdot \sigma_h$$

where

$\bar{h}$ = average home team points

σ_h = standard deviation of home team points

Away points are estimated in the same manner but using the team ratings for the away points model. This is:

$$a = d_0 + d_h - d_a$$

$$a^* = \frac{e^h}{1 + e^h}$$

Then

$$z = norminv(a^*)$$

And finally, home team estimated points are calculated as:

$$Away\ Team\ Points = \bar{a} + z \cdot \sigma_a$$

Estimating Spread

The estimated home team victory margin is computed directly from the home team points and away team points as follows:

$$Est.\ Spread = Home\ Team\ Points - Away\ Team\ Points$$

Estimating Probability of Winning

The corresponding probability of winning is determined by performing a regression analysis of actual spread as a function of estimated spread. The model has form:

$$Actual\ Spread = b_0 + b_1 \cdot Est.\ Spread$$

The solution to this model is in Table 6.7:

Table 6.7 Logit Points Model: 2015 NFL

Statistic	Home	Away
$\bar{s}$ (Average score)	23.532	21.933
σ_s (Score standard deviation)	10.176	9.432
b_0 (Home field advantage)	0.034	0.012
R^2	34.67%	
F-value	140.603	
Standard Error	11.501	

A graphical illustration of actual home team spreads compared to estimated home team spreads is in Fig. 6.9.

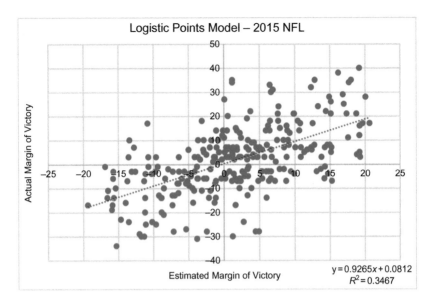

Figure 6.9 Logit Points Model: 2015 NFL.

Performance

The logit points model won 191 times out of 267 games, a winning percentage of 71.5%. When the probability of the favorite winning was over 90%, it was correct 34 times out of 35. It missed only the Eagles' 35−28 upset over the Patriots in Foxboro in Week 13, a game in which the model favored New England by 15.3 points.

This model correctly predicted 8 of the 11 playoff games, including the four division games and the two conference championships. In Super Bowl 50, the logit points model had the Panthers as a narrow 1.7-point favorite with a win probability of 56%.

Example

To determine the probability and margin of winning, we must first calculate the estimated scores for each team. The 2015 Cowboys' ratings were $b_h = 1.6343$ and $d_h = 2.9357$, while the Raiders' values were $b_a = 1.6525$ and $d_a = 1.6111$. The home field advantage factors were $b_0 = 0.0336$ and $d_0 = 0.0116$.

$$Home\ Score = 23.63$$

$$Away\ Score = 29.60$$

Table 6.8 Example Results

Model	Favorite	Underdog	Line	P(DAL Win)	P(OAK Win)
Game Scores	Oakland	Dallas	1.6	44.6%	55.4%
Team Statistics	Oakland	Dallas	4.1	36.9%	63.1%
Logistic Probability	Oakland	Dallas	7.0	25.3%	74.7%
Team Ratings	Oakland	Dallas	5.7	31.7%	68.3%
Logit Spread	Oakland	Dallas	4.2	35.8%	64.2%
Logit Points	Oakland	Dallas	6.0	30.2%	69.8%
Average	Oakland	Dallas	4.7	34.1%	65.9%

This gives us a projection that the Raiders should win by nearly 6 points:

$$Victory\ Margin = 23.63 - 29.60 = -5.97$$

The normal cumulative distribution function yields the probability of a Dallas win, given the regression's standard deviation of 11.501:

$$Probability = 1 - normcdf(0, -5.97, 11.501) = 30.19\%$$

6.7 EXAMPLE

All six models favored Oakland over the Cowboys in Dallas in our imagined matchup with odds of almost 2 to 1. The average estimated winning margin was Oakland by 4.7 points with a winning probability of 65.9%. These results are shown in Table 6.8.

6.8 OUT-SAMPLE RESULTS

We next performed an out-sample analysis where we predicted our game results using a walk forward approach. In this analysis, we use previous game results data to predict future games. Here, the model parameters were estimated after each week of games (beginning in the fifth week) and then we predicted the winning team for the next week. For all models we found the predictive power of the model declining slightly, but after 10 weeks the results from the out-sample began to converge to the results with in-sample data.

The results from our analysis showed that the game scores and team statistics models had the greatest reduction in predictive power. The game scores model had a winning percentage of about 65% but the team statistics model had a winning percentage of 62% (which was below the Vegas line of 63.7%).

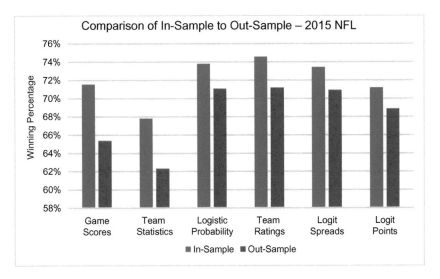

Figure 6.10 Comparison of In-Sample to Out-Sample: 2015 NFL.

The probability models also had a slight reduction in predictive power but still outperformed the Vegas line on a walk forward basis. These models all have a winning percentage of greater than 70% except for the logit points model, which had a winning percentage of 68.8%. From Week 12 through the regular season and into the postseason, the out-sample results were on par with the in-sample results, thus indicating that once we have 10−12 weeks of data the model results are as strong as they can be (Fig. 6.10).

6.9 CONCLUSION

In this chapter we applied six different sports model approaches to National Football League results for the 2015 season. The models were used to predict winning team, estimated home team victory margin, and probability of winning. We found that in all cases, these models performed better than the Vegas line using in-sample data and five of six of the models performed better than the Vegas line using out-sample data. The family of probability models (logistic probability, team ratings, logit spreads, and logit points) performed the best (highest predictive power) and outperformed the data-driven models (game scores and team statistics). The best-performing models were the logistic probability model and the team ratings model. In all cases, our modeling approaches proved to be a valuable predictor of future game results including predicting winning team, winning spread, and probability of winning.

CHAPTER 7

Basketball—NBA

This chapter applies the different sports modeling techniques to NBA basketball data for the 2014−15 season. Our goal is to provide readers with different techniques to predict the winning team, estimated victory margin, and probability of winning. These models are based on linear regression techniques described in Chapter 2, Regression Models, and on probability estimation methods described in Chapter 3, Probability Models. They include the game scores, team statistics, team ratings, logistic probability, logit spreads, and logit points models. An overview of these approaches is also provided in Chapter 5, Sports Prediction Models.

We evaluated these models using in-sample data in three different ways: winning percentage, R^2 goodness of fit, and regression error. Out-sample performance results are discussed at the end of the chapter.

Fig. 7.1 depicts the winning percentage by model. The game scores, team ratings, logistic probability, and logit spread models all picked the winners between 69.3% and 70.3% of the time. This is quite similar to the results of the 2015 NFL season with the exception of the logit points model, which by a large margin had the lowest success rate at 57.9%. The team statistics regression had the second-lowest success rate at 65.6%.

Fig. 7.2 illustrates the R^2 goodness of fit for each model. The Vegas line tied the game scores model for the best favorites' winning percentage (70.3%) and was also within 0.15 points of the lowest standard deviation for margin of victory error. Its R^2 for margin of victory of 23.5% was in line with the best-performing models.

Fig. 7.3 depicts the regression error surrounding the predicted victory margin. Here, the lower the regression error, the better the model fit. The team statistics regression also returned the largest standard deviation for the differential between the home teams' projected and actual margins of victory or defeat, 11.65 points. Next highest was the game scores model at 9.22 points. The four other models were nearly identical, ranging from 7.35 to 7.54 points. Team statistics also gave the lowest R^2 for margin of victory, 18%. The five other models had R^2 values between 22.6% and 23.9%.

Optimal Sports Math, Statistics, and Fantasy.
DOI: http://dx.doi.org/10.1016/B978-0-12-805163-4.00007-4

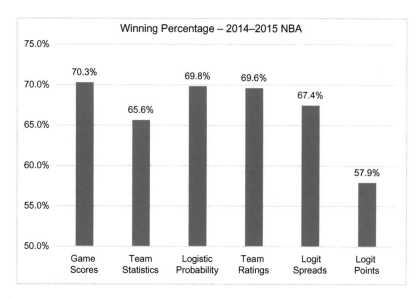

Figure 7.1 Winning Percentage: 2014–2015 NBA.

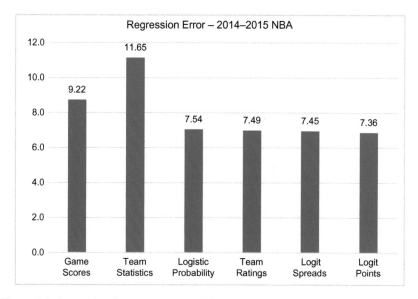

Figure 7.2 Regression Error: 2014–2015 NBA.

Table 7.1 provides the data used in these regression models to allow readers to test these models and experiment with different formulations. These models also include the rankings of the teams according to each of the six models described in the chapter.

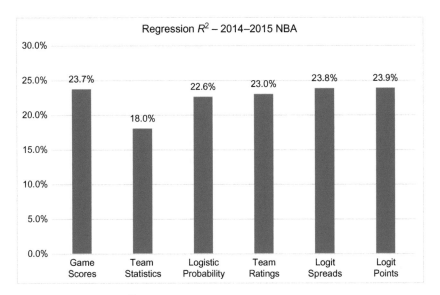

Figure 7.3 Regression R^2: 2014–2015 NBA.

7.1 GAME SCORES MODEL

The game scores regression model predicts game outcomes based on the average number of points scored and allowed by each team.

The games score model has form:

$$Y = b_0 + b_1 \cdot HPS + b_2 \cdot HPA + b_3 \cdot APS + b_4 \cdot APA + \varepsilon$$

In this representation, the dependent variable Y denotes the home team's margin of victory (or defeat). A positive value indicates the home team won by the stated number of points and a negative value indicates the home team lost by the stated number of points.

The variables of the model are:

Y = home team victory margin
HPS = home team average points scored per game
HPA = home team average points allowed per game
APS = away team average points scored per game
APA = away team average points allowed per game
b_0, b_1, b_2, b_3, b_4 = model parameters, represents the sensitivities to the corresponding model factor

The betas of this model are determined from a linear regression analysis as described in Chapter 2, Regression Models.

Table 7.1 Regression Results and Rankings

Team	W – L	Pct	Team Statistics					Ratings				Rankings				
			PF/G	PA/G	FG%	APG	RPG	Logistic Rating	Logit Spread	Logit Home Points	Logit Away Points	Game Scores	Team Statistics	Logistic Ratings	Logit Spread	Logit Points
Atlanta Hawks	60 – 22	.732	102.5	97.1	.466	25.7	40.6	3.390	2.994	2.664	2.059	4	7	5	6	5
Boston Celtics	40 – 42	.488	101.4	101.2	.443	24.5	43.8	2.299	2.430	2.348	2.442	16	14	19	17	18
Brooklyn Nets	38 – 44	.463	98.0	100.9	.451	20.9	42.4	2.307	2.066	2.379	2.842	22	20	18	24	22
Charlotte Hornets	33 – 49	.402	94.2	97.3	.420	20.2	44.1	1.985	2.006	2.207	2.774	23	28	22	25	25
Chicago Bulls	50 – 32	.610	100.8	97.8	.442	21.7	45.7	2.929	2.913	2.722	2.267	11	11	10	10	10
Cleveland Cavaliers	53 – 29	.646	103.1	98.7	.458	22.1	43.0	3.353	3.158	2.826	2.062	5	10	6	4	4
Dallas Mavericks	50 – 32	.610	105.2	102.3	.463	22.5	42.3	3.041	2.951	2.884	2.339	10	9	9	9	9
Denver Nuggets	30 – 52	.366	101.5	105.0	.433	21.8	44.7	1.909	2.075	2.366	2.862	24	21	24	23	24
Detroit Pistons	32 – 50	.390	98.5	99.5	.432	21.6	44.9	1.922	2.341	2.702	2.910	20	22	23	20	20
Golden State Warriors	67 – 15	.817	110.0	99.9	.478	27.4	44.7	4.254	3.857	3.327	1.740	1	1	1	1	1
Houston Rockets	56 – 26	.683	103.9	100.5	.444	22.2	43.7	3.479	2.970	2.610	2.024	7	17	2	8	8
Indiana Pacers	38 – 44	.463	97.3	97.0	.439	21.4	44.9	2.253	2.474	2.494	2.525	17	16	20	16	16
Los Angeles Clippers	56 – 26	.683	106.7	100.1	.473	24.8	42.6	3.431	3.399	2.877	1.869	2	4	3	2	2
Los Angeles Lakers	21 – 61	.256	98.5	105.3	.435	20.9	43.9	1.355	1.695	1.894	2.872	27	25	27	27	27
Memphis Grizzlies	55 – 27	.671	98.3	95.1	.458	21.7	42.6	3.430	2.991	2.717	2.116	9	15	4	7	6
Miami Heat	37 – 45	.451	94.7	97.3	.456	19.8	39.1	2.214	2.090	2.478	2.922	21	26	21	22	21
Milwaukee Bucks	41 – 41	.500	97.8	97.4	.459	23.6	42.1	2.405	2.355	2.405	2.507	15	13	17	19	19
Minnesota Timberwolves	16 – 66	.195	97.8	106.5	.438	21.6	40.9	0.984	1.444	1.642	2.884	28	27	29	28	28
New Orleans Pelicans	45 – 37	.549	99.4	98.6	.457	22.0	43.5	2.733	2.656	2.738	2.560	13	8	14	13	13
New York Knicks	17 – 65	.207	91.9	101.2	.428	21.3	40.4	0.932	1.226	1.648	3.072	30	29	30	30	30
Oklahoma City Thunder	45 – 37	.549	104.0	101.8	.447	20.5	47.5	2.769	2.793	2.773	2.384	12	5	12	11	11
Orlando Magic	25 – 57	.305	95.7	101.4	.453	20.6	41.8	1.520	1.723	2.226	3.100	26	24	26	26	26
Philadelphia 76ers	18 – 64	.220	92.0	101.0	.408	20.5	42.9	1.034	1.245	2.015	3.384	29	30	28	29	29
Phoenix Suns	39 – 43	.476	102.4	103.3	.452	20.2	43.2	2.454	2.427	2.333	2.411	19	18	15	18	17
Portland Trail Blazers	51 – 31	.622	102.8	98.6	.450	21.9	45.9	3.047	3.023	2.737	2.137	6	6	8	5	7
Sacramento Kings	29 – 53	.354	101.3	105.0	.455	20.3	44.2	1.882	2.116	2.070	2.536	25	12	25	21	23
San Antonio Spurs	55 – 27	.671	103.2	97.0	.468	24.4	43.6	3.341	3.348	3.259	2.287	3	2	7	3	3
Toronto Raptors	49 – 33	.598	104.0	100.9	.455	20.7	41.5	2.753	2.722	2.351	2.094	8	23	13	12	12
Utah Jazz	38 – 44	.463	95.1	94.9	.447	19.9	44.0	2.407	2.602	2.531	2.387	18	19	16	14	14
Washington Wizards	46 – 36	.561	98.5	97.8	.462	24.0	44.7	2.800	2.588	2.763	2.623	14	3	11	15	15
Home-Court Advantage								0.387	0.321	0.013	0.011					

The probability that the home team will be victorious is computed from the Excel normal distribution function as follows:

$$p = 1 - NormDist(0, Y, SeY, True)$$

Here, Y and SeY are the expected victory margin and regression error term respectively, and zero indicates the reference point used for the calculation. That is, the probability that the winning margin will be greater than zero.

Regression Results

The best fit regression equation for predicting the victory margin as a function of team points scored and points allowed is below. The resulting victory margin is positive when the home team is favored and negative when the visiting team is favored.

This model equation is:

$$Est.\ Victory\ Margin = -1.28 + 0.867 \cdot HPS - 0.823 \cdot HPA$$
$$- 0.609 \cdot APS + 0.999 \cdot APA$$

The game scores model yielded a moderately high R^2 value of 23.7%. The t-Stats were -11.68 and 7.74 for visiting teams' points scored and points allowed, and 13.72 and -8.89 for home teams, and an F-value of 101.55.

The signs of the input variables are intuitive. The sign of the sensitivity parameter is expected to be positive for home points scored (HPS) and away points allowed (APA). In both of these cases, the home team will score more points if these input values are higher. Similarly, the signs of the sensitivity parameters for home points allowed (HPA) and away team points scored (APS) are expected to be negative. The home team is expected to win by fewer points or possibly lose if the HPA and/or APS increase in value.

A graph showing the actual spreads as a function of estimates spreads is shown in Fig. 7.4 and Table 7.2.

Performance

The game scores regression accurately identified the winner 921 times in 1311 games (70.25%). The probability predictions had a very high goodness of fit of 86.3% when compared with teams' actual winning percentages at a granularity of 5 percentage points, and the t-statistic was 7.09. When the favorite's probability of winning was over 60%, the

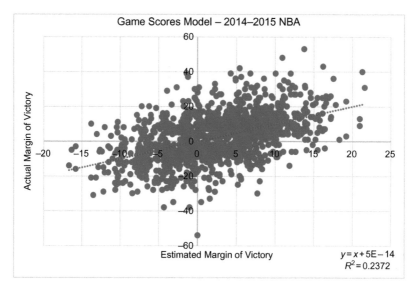

Figure 7.4 Game Scores Model: 2014–2015 NBA.

Table 7.2 Game Scores Model: 2014–2015 NBA		
Statistic	**Value**	***t*-Stat**
b_0	−0.442	0.0
b_1 (Home points scored)	1.068	13.7
b_2 (Home points allowed)	−1.008	−8.9
b_3 (Visitor points scored)	0.078	−11.7
b_4 (Visitor points allowed)	0.878	7.7
R^2	23.72%	
F-value	101.546	
Standard Error	11.821	

favorite won 75.8% of the time. The correct prediction rate climbed to 82.3% above 70% confidence, and 90% correct above 80% confidence. Above 90% confidence, the prediction was wrong only twice out of 37 games (94.6%), both of which were road wins by the Knicks. The first was a five-point victory over the Cleveland Cavaliers on October 30, 2014, when LeBron James, in his first home game since returning to the Cavaliers, shot 5 of 15 from the floor to finish with just 17 points and 8 turnovers. The other came on April 13, 2015, against the Atlanta Hawks, who had long since sewn up the top seed in the Eastern Conference, in what was each team's second-to-last regular-season game.

Predictions for postseason series fell into one of two categories. If the difference between the teams was smaller than the home-court advantage factor, the model would favor the home team in each game. If one team was stronger than the other by more than the home-court advantage factor, that team would be favored to win every game. The Golden State Warriors were the clear favorite in each postseason series, as they went 16—5 in the playoffs en route to the NBA Championship. There were clear favorites in three of the four Eastern Conference first-round series (Hawks, Bulls, Cavaliers), but none in the Western Conference other than Golden State. The only other series in which there was a clear favorite was the Eastern Conference semifinals between the Hawks and the Washington Wizards. Of the 15 playoff series, there were clear favorites in eight, and the favorite did indeed win all eight. For the other seven series, the favored home team compiled a win—loss record of 23—16 (.590).

The Vegas line predicted the winner 912 times out of 1297 games for a winning percentage of 71.7%, excluding 14 pick 'em games where Vegas did not name a favorite.

Rankings

The game scores model's top ranking went to the champion Warriors (rating 29.195), with the Los Angeles Clippers (22.274) and San Antonio Spurs (21.212) coming in second and third. The Clippers and Spurs met in the first round of the playoffs; Los Angeles prevailed in the only postseason series that went the full seven games. The two Eastern Conference finalists, the Hawks and the Cavaliers, came in at fourth and fifth. Fifteen of the top sixteen teams made the postseason. The highest-ranked nonplayoff team was the Oklahoma City Thunder at No. 12; they had lost the tiebreaker with the New Orleans Pelicans for the Western Conference's last playoff spot. The lowest-ranked playoff team was the No. 22 Brooklyn Nets, who put up a regular season record of just 38—44 playing in the NBA's weakest division. The Atlantic Division was the only one of the six that produced no 50-game winners, but it was home to not one but two teams that failed to win 20. The Eastern Conference's three divisions accounted for the same number of teams that finished over .500 (five) as the Western Conference's Southwest Division did by itself. Brooklyn was ranked lower than three nonplayoff Eastern Conference teams: the Indiana Pacers (No. 17, against whom the Nets also won the tiebreaker), the Detroit Pistons (No. 20), and the Miami Heat (No. 21).

Example

To compare the six models, we will examine a matchup of two top-tier Western Conference teams, the Golden State Warriors (67−15) and the San Antonio Spurs (55−27). To make things a little closer, let's make this a road game for the more powerful Warriors.

The 2014−15 Spurs scored 103.2 points per game while allowing 97.0. The Warriors' offense averaged 110.0 points while their defense gave up 99.9 points per game. Given those averages and the betas from the regression, the game scores model favors the Warriors in this game by the extremely slim margin of 0.4 points:

$$Est.\ Victory\ Margin = -0.44 + 1.07(103.2) - 1.01(97.0) - 0.91(110.0)$$
$$+0.88(99.9) = -0.40 \pm 11.8$$

The corresponding probability that the host Spurs would win is 48.65%:

$$p = 1 - NormCDF(0, -0.40, 11.82) = 48.65\%$$

7.2 TEAM STATISTICS MODEL

The team statistics regression uses team performance statistics to predict game results. Generally speaking, these measurements could be either per-game averages (such as rebounds per game) or per-event averages (such as field goal shooting percentage). Proper team performance statistics should also encompass both offensive and defensive ability. Through experimentation, readers can determine which set of team statistics provides the greatest predictive power.

In this section, we will demonstrate the regression using three statistics for each team: field goal shooting percentage, assists per game, and total rebounds per game.

The team statistics linear regression model has form:

$$Y = b_0 + b_1 \cdot HTFG\% + b_2 \cdot HTAPG + b_3 \cdot HTRPG + b_4 \cdot ATFG\%$$
$$+ b_5 \cdot ATAPG + b_6 \cdot ATRPG + \varepsilon$$

The variables of the model are:

Y = outcome value we are looking to predict (spread, home team points, away team points, and total points)

$TFG\%$ = home team field goal percentage

$HTAPG$ = home team assists per game

$HTRPG$ = home team total rebounds per game
$ATFG\%$ = away team field goal percentage
$ATAPG$ = away team assists per game
$ATRPG$ = away team total rebounds per game
$b_0, b_1, b_2, b_3, b_4, b_5, b_6$ = model parameter, sensitivity to the variable
ε = *model error*

The betas of this model are determined from a linear regression analysis as described in Chapter 2, Regression Models. The results of this model are shown in Table 7.3.

Table 7.3 Team Statistics Model: 2014−2015 NBA		
Statistic	**Value**	**t-Stat**
b_0	−38.553	−1.9
b_1 (Home field goal percentage)	215.048	7.5
b_2 (Home assists per game)	0.587	2.5
b_3 (Home rebounds per game)	1.362	6.9
b_4 (Away field goal percentage)	−198.323	−6.9
b_5 (Away assists per game)	−0.408	−1.8
b_6 (Away rebounds per game)	−0.682	−3.5
R^2	18.04%	
F-value	47.823	
Standard Error	12.263	

Regression Results

The result of the regression model for predicting home team victory margin from our team statistics is:

$$Est.\,Victory\,Margin = -38.553 + 215.048(HTFG\%) + 0.587(HTAPG)$$
$$+ 1.362(HTRPG) - 198.323(ATFG\%)$$
$$- 0.408(ATAPG) - 0.682(ATRPG) \pm 12.263$$

The model produced an R^2 of 18.0% and a standard error of 12.26 points. The t-statistic for the road team's free throw percentage was not significant at −1.77, while that of the home team's field goal percentage was just over 7.5.

Actual victory margins with respect to the home team ranged from −54 (Bulls 120, Bucks 66 in the first round of the playoffs) to +53 (Mavericks 123, 76ers 70 in November) with a standard deviation of

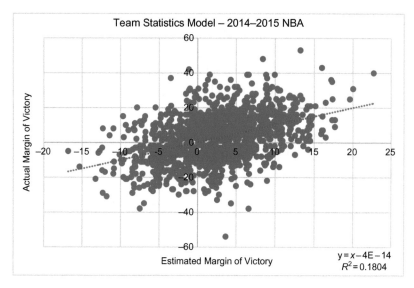

Figure 7.5 Team Statistics Model: 2014–2015 NBA.

13.5 points. The projected home team victory margins ranged from −17.84 to 20.21 points, a standard deviation of 5.74 points. Differences between projected and actual victory margins had a standard deviation of 12.2 points. The model was accurate to within one point 81 times (6.2%), three points in 254 games (19.4%), and six points 501 times (38.2%). It was off by 20 points or more only 9.4% of the time.

These regression results and a graph of actual spread compared to predicted spread is shown in Fig. 7.5 and Table 7.3.

Performance

The team statistics model correctly predicted the winner in 860 of 1311 games, a success rate of 65.6%. The model was only 40.5% accurate when the favorite was estimated as having a win probability between 50% and 55%, but that was the only interval for which it underperformed. When the model calculated at least an 85% chance of winning, the favorite won 55 times and lost only twice (96.5%); one was the Warriors' two-point loss at home to the Bulls in late January, and the other a four-point road win by the Hornets over the Wizards the following week. Above 87.7% it was a perfect 34 for 34. Overall the goodness of fit between the estimated win probability and actual win percentage was 95.8%.

The Vegas model predicted the winner 912 times out of 1297 games for a winning percentage of 71.7%, excluding 14 pick 'em games where Vegas did not name a favorite.

Rankings

Unsurprisingly the Golden State Warriors finished atop the team statistics model's rankings, as teams from the Western Conference accounted for five of the top six spots. San Antonio, Washington, and the Los Angeles Clippers finished close together in the No. 2 through No. 4 slots. Oklahoma City, who lost the tiebreaker for the last playoff berth in the West, was ranked No. 5, while Sacramento (12) and Indiana (16) were other nonplayoff teams in the top 16. Brooklyn was the lowest-ranked playoff team at No. 20, behind Utah and Phoenix. The five teams at the bottom of the rankings included three that won no more than 18 games—the Timberwolves, Knicks, and 76ers—and two that won at least 33, the No. 26 Heat (37−45) and the No. 28 Hornets (33−49).

Example

The team statistics model comes to a nearly identical conclusion as the game scores model. Golden State has slight edges in all three categories: field goal percentage (.488 to .478), rebounding (44.7 to 43.6), and assists per game (27.4 to 24.4). The Warriors are still favored, but owing to the home-court advantage factor the margin is still less than 1 point:

$$
\begin{aligned}
Est\ Margin\ of\ Victory = {} & -38.553 + 215.048(.468) + 0.587(24.4) \\
& + 1.362(43.6) - 198.323(.478) - 0.408(27.4) \\
& - 0.682(44.7) \pm SE = -0.57 \pm 12.26
\end{aligned}
$$

The margin of −0.57 points is equivalent to a win probability of 48.1% for Golden State:

$$
p = 1 - NormCDF(0, -0.57, 12.26) = 48.1\%
$$

7.3 LOGISTIC PROBABILITY MODEL

The logistic probability model infers a team strength rating based only on game outcomes such as whether the team won, lost, or tied the game. The result of the game is determined from the perspective of the home team, but analysts can use the same approach from the perspective of the visiting team.

The logistic model is as follows:

$$\gamma^* = \frac{1}{1 + \exp\left\{-(b_0 + b_h - b_a)\right\}}$$

Here, b_0 denotes a home-field advantage parameter, b_h denotes the home team rating parameter value, and b_a denotes the away team rating parameter value. The value γ^* denotes the probability that the home team will win the game. Team ratings for the logistic probability are determined via maximum likelihood estimates and are shown in Table 7.1.

Estimating Spread

The estimated spread (i.e., home team victory margin) is determined via a second analysis where we regress the actual home team spread on the estimated probability γ^* (as the input variable). This regression has form:

$$Actual\ Spread = a_0 + a_1 \cdot \gamma^*$$

This model now provides a relationship between the logistic home team winning probability and the home team winning percentage. It is important to note here that analysts may need to incorporate an adjustment to the spread calculation if the data results are skewed; (see Chapter 3: Probability Models).

The solution to this model is (Table 7.4):

Table 7.4 Logistic Probability Regression: 2014–2015 NBA

Statistic	Value	t-Stat
b_0	−14.003	−15.6
b_1	28.466	19.6
b_0 (Home-court advantage)	0.387	
R^2	22.61%	
F-value	382.473	
Standard Error	11.893	

After computing the home team winning probability, the expected spread is estimated from the following equation based on the regression results:

$$Estimated\ Spread = -14.003 + 28.466 \cdot \gamma^*$$

A graph illustrating the estimated spread from the probability estimates is shown in Fig. 7.6.

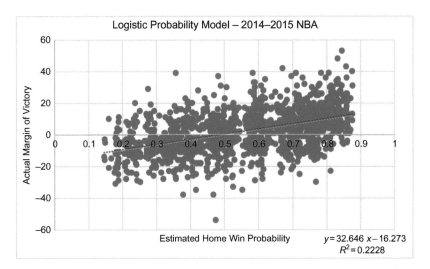

Figure 7.6 Logistic Probability Model: 2014–2015 NBA.

Performance

The logistic probability model had an accuracy of 69.8% for selecting the winner, as the teams it favored won 915 of 1311 games over the regular season and the playoffs. The logistic probability model also proved to be a very good predictor of the victory margin with an $R^2 = 22.6\%$ and standard error of 11.89. The standard error of the Vegas line was 11.82.

The model's favorites won 741 of 984 games with projected win probabilities of 60% or higher, a rate of 75.3%. When the win probability was above 85%, their win rate rose to 91.1%. The largest upset was the Knicks' road win in Atlanta in April; the model gave the Hawks a 94.5% chance of winning, making New York a 17-to-1 underdog. The home team was the favorite 61.7% of the time.

Rankings

The top of the logistic rankings was dominated by teams from the Western Conference, led by the eventual NBA Champion Golden State Warriors and followed by the Houston Rockets, Los Angeles Clippers, and Memphis Grizzlies. Only two of the top nine teams came from the Eastern Conference, the Atlanta Hawks at No. 5 and the Cleveland Cavaliers at No. 6.

Teams bound for the playoffs made up the top 11 places in the logistic rankings. The Oklahoma City Thunder, who did not make the playoffs, was ranked at No. 12, ahead of five teams who did, including the No. 13

Toronto Raptors, the No. 4 seed in the Eastern Conference. In the bottom half of the rankings, Eastern Conference teams outnumbered their Western counterparts 2 to 1.

Example

The Warriors' logistic rating of 4.2544 led the NBA, while San Antonio came in seventh with 3.3412. So far the game scores and team statistics models have predicted an extremely narrow win by Golden State. The logistic probability model favors the Warriors by a somewhat larger margin:

$$y^* = \frac{1}{1 + e^{-(0.3871 + 3.3412 - 4.2544)}} = \frac{1}{1 + e^{0.5261}} = 0.3714$$

We can then take this value and combine it with the regression parameters b_0 (-14.003) and b_1 (28.466) to estimate that this matchup should result in a Golden State win by nearly 3½ points, and win probabilities of 38.65% for the Spurs and 61.35% for the Warriors:

$$Estimated\ Spread = -14.003 + 28.466 \cdot 0.3714 \pm SE = -3.43 \pm 11.89$$

$$Probability = 1 - normdist(0, -3.43, 11.89) = 38.65\%$$

7.4 TEAM RATINGS MODEL

The team ratings prediction model is a linear regression model that uses the team ratings determined from the logistic probability model as the explanatory variables to estimate home team victory margin. This is one of the reasons why the logistic model is among the more important sports models since its results can be used in different modeling applications.

The team ratings regression has form:

$$Y = b_0 + b_1 \cdot Home\ Rating + b_2 \cdot Away\ Rating$$

The variables of the model are:

Home Rating = home team strength rating (from logistic probability model)

Away Rating = away team strength rating (from logistic probability model)

b_0 = home-court advantage value

b_1 = home team rating parameter

b_2 = away team rating parameter

Y = value we are looking to predict

The probability of winning is determined from:

$$Probability = 1 - NormCDF(0, Y, SE)$$

The betas of this model, b_1 and b_2, are determined from a linear regression analysis as described in Chapter 2, Regression Models.

Regression Results

The best fit regression equation to predict home team victory margin from team strength ratings is: (Fig. 7.7, Table 7.5).

$$Victory\ Margin = 0.9925 + 5.5642 \cdot Home\ Rating - 5.2037 \cdot Away\ Rating$$

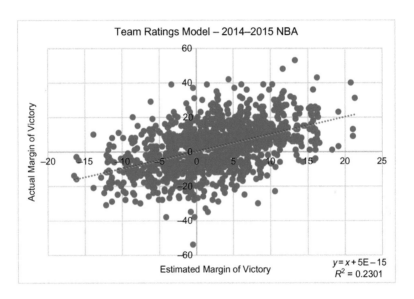

Figure 7.7 Team Ratings Model: 2014–2015 NBA.

Table 7.5 Team Ratings Model: 2014–2015 NBA		
Statistic	**Value**	**t-Stat**
b_0	− 0.016	0.0
b_1 (Home team rating)	6.159	15.4
b_2 (Away team rating)	− 5.213	− 13.0
R^2	23.01%	
F-value	195.424	
Standard Error	11.867	

Performance

This model's expected favorites won 912 of 1311 regular and postseason NBA games for 2014−15, a rate of 69.6%. When the estimated probability of Team A's defeating Team B was at least 75%, A did go on to win 321 times out of 365 (87.9%).

The two biggest upsets according to the ratings regression were the same two as in the game scores regression, i.e., the Knicks' two road wins against the Cavaliers (9.2% chance the Knicks would win) in October and in Atlanta against the Hawks in April (9.1%). There were only four other upsets by underdogs with a win probability under 15%, all of which belonged to the Lakers when they defeated the Warriors, Rockets, Hawks, and Spurs.

The Vegas model predicted the winner 912 times out of 1297 games for a winning percentage of 71.7%, excluding 14 pick 'em games where Vegas did not name a favorite. The ratings regression predicted the winner 903 times out of 1311 games for a winning percentage of 68.9%.

Rankings

The 16 playoff teams all finished in the top 19 of the rankings. Golden State once again took the top spot, and owing to their dominance the next three spots were also filled by Western Conference teams. The Eastern Conference managed only two teams in the top nine, the Hawks at No. 5 and the Cavaliers at No. 6. The highest-ranked nonplayoff team was the Oklahoma City Thunder at No. 12, who lost the tiebreaker for the last playoff berth in the West to New Orleans, two spots below. Phoenix (No. 15) and Utah (No. 17) also finished ahead of the bottom two Eastern Conference playoff teams, the Nets (No. 18) and the Celtics (No. 19).

Example

The ratings regression also predicts a close game by very nearly the same score, but this time the projected margin of victory is wider at a little over a point and a half. The Warriors came out on top of the rakings with a rating of 4.254, while San Antonio finished in the No. 7 spot with a rating of 3.341.

$$Est.\ Victory\ Margin = -0.016 + 6.159(3.341) - 5.213(4.254) \pm SE$$
$$= -1.62 \pm 11.87$$

The estimated winning probability is 44.6% for San Antonio and 55.4% for Golden State:

$$Probability = 1 - NormCDF(0, -1.62, 11.87) = 44.6\%$$

7.5 LOGIT SPREAD MODEL

The logit spread model is a probability model that predicts the home team victory margin based on an inferred team rating metric. The model transforms the home team victory margin to a probability value between zero and one via the cumulative distribution function and then estimates model parameters via logit regression analysis.

The logit spread model has following form:

$$y^* = b_0 + b_h - b_a$$

where b_0 denotes a home-field advantage parameter, b_h denotes the home team parameter value, and b_a denotes the away team parameter value.

The left-hand side of the equation y is the log ratio of the cumulative density function of victory margin (see chapter: Sports Prediction Models, for calculation process).

In this formulation, the parameters values b_0, b_h, b_a are then determined via ordinary least squares regression analysis. The results of this analysis are shown in Table 7.6.

Estimating Spreads

Estimating the home team winning margin is accomplished as follows.

If team k is the home team and team j is the away team, we compute y using the logit parameters:

$$y = b_0 + b_k - b_j$$

Compute y^* from y via the following adjustment:

$$y^* = \frac{e^y}{1 + e^y}$$

Compute z as follows:

$$z = norminv(y^*)$$

And finally, the estimated home team spread:

$$Estimated\ Spread = \bar{s} + z \cdot \sigma_s$$

where

> $\bar{s}$ = average home team winning margin (spread) across all games
> σ_s = standard deviation of winning margin across all games

Estimating Probability

The corresponding probability of the home team winning is determined by performing a regression analysis of actual spread as a function of estimated spread. The model has form:

$$Actual\ Spread = b_0 + b_1 \cdot spread$$

The solution to this model is (Table 7.6):

Table 7.6 Logit Spread Model: 2014–2015 NBA	
Statistic	**Value**
$\bar{s}$ (Average home victory margin)	2.391
σ_s (Home victory margin standard deviation)	13.514
b_0 (Home-court advantage)	0.321
R^2	23.84%
F-value	409.749
Standard Error	11.798

A graphical illustration of this model is (Fig. 7.8):

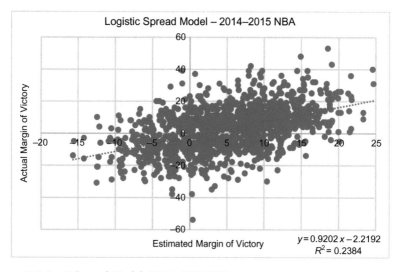

Figure 7.8 Logit Spread Model: 2014–2015 NBA.

Performance

The logit spread model correctly identified the winner in 884 of 1311 games, a rate of 67.4%, which is the second-most accurate of the six. When the favorite's probability of winning was over 75%, the favorite won 87% of the time. There were only two upsets among the 47 games where one team had a win probability of 88% or better, the aforementioned Knicks wins over Atlanta (90.0%) and Cleveland (90.5%).

Rankings

Golden State again finished atop the ratings for the logit spread model, followed by the Clippers and the Spurs. The Hawks, Cavaliers, and Bulls, the top three seeds in the Eastern Conference, placed sixth, fourth, and tenth respectively. The Thunder again was the highest-ranked team that did not qualify for postseason play. The No. 8-seed in the East, the Brooklyn Nets, ranked 24th, behind even three teams (the Detroit Pistons, Sacramento Kings, and Denver Nuggets) with winning percentages below .400.

Example

The Warriors' logit spread parameter of 3.8573 again led the NBA, while the Spurs' value of 3.2586 was good enough for third out of the 30 teams. The home-court advantage factor (b_0) was 0.3207 for the NBA in 2015.

$$y = \frac{1}{1 + e^{-(0.321 + 3.8537 - 4.254)}} = \frac{1}{1 + e^{-0.592}} = 0.453$$

The average NBA game in 2014—15 ended with the home team winning by 2.391 points with a standard deviation of 13.514. The logit spread expects San Antonio to win by less than one point:

Victory Margin $= z \cdot \sigma_s + \bar{s} = norminv(0.453) \cdot 13.514 + 2.391 = 0.796$

With an estimated margin of victory of 0.796 points, the model puts the win probability at 52.69% for San Antonio and 47.31% for Golden State:

Probability $= 1 - normdist(0, 0.796, 11.798) = 52.69\%$

7.6 LOGIT POINTS MODEL

The logit points model is a probability model that predicts the home team victory margin by taking the difference between home team predicted points and away team predicted points. The predicted points are determined based on inferred team "ratings" similar to the logit spread model discussed above.

The logit points model has following form:

$$h^* = b_0 + b_h - b_a$$
$$a^* = d_0 + d_h - d_a$$

where

> h is the transformed home team points, c_0 denotes a home-field advantage, c_h denotes the home team rating, and c_a denotes the away team rating corresponding to home team points
>
> a is the transformed away team points, d_0 denotes a home-field advantage, d_h denotes the home team rating, d_a denotes the away team rating corresponding to away team points

The left-hand side of the equation h and a is the log ratio of the cumulative density function of home team points and away team points respectively (see chapter: Sports Prediction Models, for a description).

Estimating Home and Away Team Points

Estimating the home team points is accomplished directly from the home points team ratings. These rating parameters are shown in Table 7.1. If team k is the home team and team j is the away team, the transformed home team points is:

$$h = c_0 + c_k - c_j$$

$$h^* = \frac{e^h}{1 + e^h}$$

Then,

$$z = norminv(h^*)$$

And finally, the x value is:

$$Home\ Points = \overline{h} + z \cdot \sigma_h$$

where

> $\overline{h}$ = average home team points
>
> σ_h = standard deviation of home team points

Away points are estimated in the same manner but using the team ratings for the away points model.

Estimating Spread

The estimated home team victory margin is computed directly from the home team points and away team points as follows:

$$Est.\ Spread = Home\ Team\ Points - Away\ Team\ Points$$

Estimating Probability of Winning

The corresponding probability of winning is determined by performing a regression analysis of actual spread as a function of estimated spread. The model has form:

$$Actual\ Spread = b_0 + b_1 \cdot Est.\ Spread$$

The solution to this model is: (Table 7.7).

Table 7.7 Logit Points Model: 2014–2015 NBA

Statistic	Home	Away
$\bar{s}$ (Average score)	101.243	98.852
σ_s (Score standard deviation)	11.690	11.725
b_0 (Home-court advantage)	0.013	0.011
R^2		23.87%
F-value		410.365
Standard Error		11.796

A graphical illustration of this model is: (Fig. 7.9).

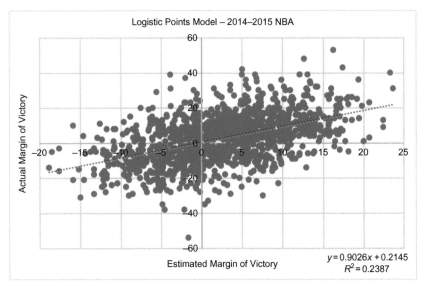

Figure 7.9 Logit Points Model: 2014–2015 NBA.

Performance

The logit points model's favorite won 907 times out of 1311 games, a winning percentage of 69.2%. When the spread was at least 11 points, the favorite won 90% of the time (177 out of 197). Above six points, the model was correct in 465 times out of 570 (81.6%). The two biggest upsets were again the Knicks' road wins over the Cavaliers (Cleveland was favored by 18.1 points) and the Hawks (Atlanta was favored by 17.03 points). The spreads were within ± 1 point of the actual margins of victory 91 times (6.94%), within ± 3 points 283 times (21.6%), and within ± 6 points 537 times (41.0%).

Rankings

The Warriors were ranked No. 1 by the logit points model, as Golden State took the top ranking in each of the six models. The top nine spots were comprised of seven playoff teams from the Western Conference and two from the East. The West also contributed the only nonplayoff teams in the top 50%, the Oklahoma City Thunder (No. 11) and the Utah Jazz (No. 14). Utah finished one spot above the Washington Wizards, the 5-seed in the East, despite having eight fewer wins, in another demonstration of the relative dominance of the Western Conference.

Example

San Antonio had values of $b_h = 3.2586$ and $d_h = 2.2867$, and those for Golden State were $b_a = 3.3268$ and $d_a = 1.7399$. The home-court advantage factors were $b_0 = 0.0133$ and $d_0 = 0.0109$.

$$Home\ Score = norminv\left(\frac{1}{1 + e^{-(0.0133 + 3.2586 - 3.327)}}\right) \cdot 11.6987 + 101.2426$$

$$= 100.84$$

$$Away\ Score = norminv\left(\frac{1}{1 + e^{-(0.0109 + 2.2867 - 1.7399)}}\right) \cdot 11.7247 + 98.8520$$

$$= 102.93$$

Subtracting the away score from the home score results in a margin of just over 2 points in favor of the visiting Warriors:

$$Victory\ Margin = 100.84 - 102.93 = -2.09$$

Taking the normal cumulative distribution of this margin and the regression's standard deviation of 11.769, the model reports that the Spurs have a 43% chance of defeating the Warriors at home:

$$Probability = 1 - normcdf(0,\ -2.09,\ 11.796) = 42.98\%$$

7.7 EXAMPLE

Overall, five of the six models favor Golden State in our imagined matchup with the Spurs in San Antonio, while the logit spread model calls this game for Gregg Popovich's team. Three models estimated that the favorite would win by less than 1 point, while only the logistic probability model gave its favorite more than a 60% probability of winning. Averaged together, Golden State has a 54.0% chance of a road win in San Antonio, with a margin of 1.2 points (Table 7.8).

7.8 OUT-SAMPLE RESULTS

We performed an out-sample analysis where we predicted our game results using a walk forward approach. In this analysis, we use previous game results data to predict future games. Here, the model parameters were estimated after about 10 games per team and then we predicted the winning team for the next game. We found the predictive power of the model only declining slightly, but after 20 games per team the out-sample began to converge to the results with in-sample data.

The results from our analysis showed that the game scores (-1.9%) and team statistics (-3.6%) models had the greatest reduction in predictive power. The game scores model had an out-sample winning percentage of about 68% but the team statistics model had an out-sample winning percentage of 62%.

The probability models only had a very slight reduction in predictive power based on a walk forward basis. These models all have a winning percentage consistent with the in-sample results and the logit points had a higher winning percentage. After about 20 games per team (i.e., one-fourth

Table 7.8 Example Results					
Model	Favorite	Underdog	Line	P(SA Win)	P(GS Win)
Game Scores	Golden State	San Antonio	0.4	48.7%	51.3%
Team Statistics	Golden State	San Antonio	0.6	48.1%	51.9%
Logistic Probability	Golden State	San Antonio	3.4	38.7%	61.3%
Team Ratings	Golden State	San Antonio	1.6	44.6%	55.4%
Logit Spread	San Antonio	Golden State	0.8	52.7%	47.3%
Logit Points	Golden State	San Antonio	2.1	43.0%	57.0%
Average	Golden State	San Antonio	1.2	45.9%	54.1%

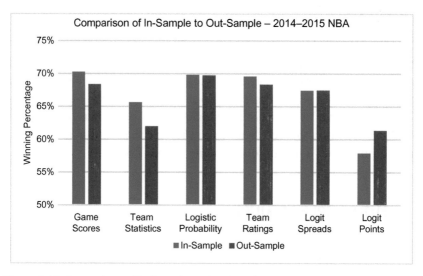

Figure 7.10 Comparison of In-Sample to Out-Sample: 2014−2015 NBA.

of the season) and into the postseason, the out-sample results were on par with the in-sample results (Fig. 7.10).

7.9 CONCLUSION

In this chapter we applied six different sports model approaches to National Basketball Association results for the 2014−15 season. The models were used to predict winning team, estimated home team victory margin, and probability of winning. Three of the six models accurately predicted the winner 70% of the time with two others better than 65%; five had regression R^2 values of between 22.6% and 23.9%. The four probability models (logistic probability, team ratings, logit spreads, and logit points) had an average regression error of 7.46 points, nearly 3 points fewer than the average of the two data-driven models (game scores and team statistics). These modeling approaches performed well in terms of predicting winning teams, winning spread, and probability of winning.

CHAPTER 8

Hockey—NHL

In this chapter, we apply the sports modeling techniques introduced in previous chapters to the 2014—15 National Hockey League (NHL) season. Our goal is to provide readers with proper techniques to predict expected winning team and probability of winning, and to estimate winning victory margin.

Our techniques include six different models: game scores, team statistics, team ratings, logistic probability, logit spreads, and logit points. These models are based on linear regression techniques described in Chapter 2, Regression Models, and on probability estimation methods described in Chapter 3, Probability Models.

The models were evaluated in three different ways based on in-sample data: winning percentage, R^2 goodness of the estimated victory margin, and the regression error. Out-sample performance results are discussed at the end of the chapter.

The six models performed very similar to each other. All had accuracy rates between 58% and 61%, and five had R^2 values between 8% and 9%. At 5.6%, only the team statistics model, using one particular set of parameters, had an R^2 below 8%. The standard deviations of the differences between expected and actual margins of victory were also nearly identical, all coming in between 2.20 and 2.25 (Figs. 8.1—8.3 and Table 8.1).

8.1 GAME SCORES MODEL

The game scores regression model predicts game outcomes based on the average number of goals scored and allowed by each team.

The game scores model has form:

$$Y = b_0 + b_1 \cdot HGS + b_2 \cdot HGA + b_3 \cdot AGS + b_4 \cdot AGA + \varepsilon$$

In the representation, the dependent variable Y denotes the value that we are trying to predict; this can be the home team's margin of victory (or defeat), the number of goals scored by the home team, the number of goals scored by the away team, and the total combined score. For our purposes,

Optimal Sports Math, Statistics, and Fantasy.
DOI: http://dx.doi.org/10.1016/B978-0-12-805163-4.00008-6

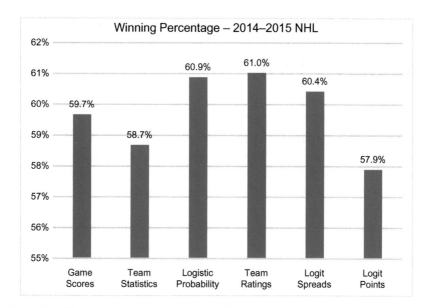

Figure 8.1 Winning Percentage: 2014–15 NHL.

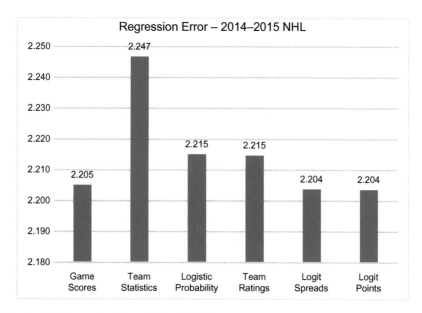

Figure 8.2 Regression Error: 2014–15 NHL.

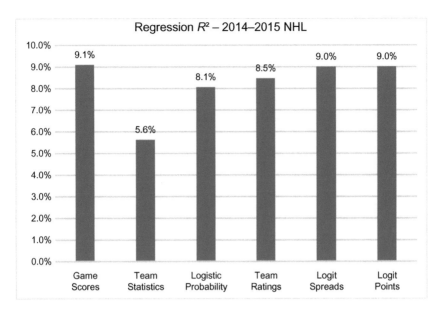

Figure 8.3 Regression R^2: 2014–15 NHL.

we define the Y variable to be the home team victory margin. A positive value indicates the home team won by the stated number of goals and a negative value indicates the home team lost by the stated number of goals.

The variables of the model are:

Y = home team victory margin
HGS = home team average goals scored per game
HGA = home team average goals allowed per game
AGS = away team average goals scored per game
AGA = away team average goals allowed per game
b_0, b_1, b_2, b_3, b_4 = model parameters, represents the sensitivities to the corresponding model factor.

The betas of this model are determined from a linear regression analysis as described in Chapter 2, Regression Models.

The probability that the home team will be victorious is computed from the Excel normal distribution function as follows:

$$p = 1 - NormDist(0, Y, SeY, True)$$

Here, Y and SeY are the expected victory margin and regression error term respectively, and zero indicates the reference point used for the calculation (i.e., the probability that the winning margin will be greater than zero).

Table 8.1 Regression Results and Rankings

Team	W – L – OTL	Team Statistics					Ratings				Rankings				
		Points	GF/G	GA/G	S%	SV%	Logistic Rating	Logit Spread	Logit Points (Home)	Logit Points (Away)	Game Scores	Team Statistics	Logistic Ratings	Logit Spread	Logit Points
Anaheim Ducks	51 – 24 – 7	109	2.8	2.7	.0928	.907	3.092	2.697	2.538	2.251	18	17	1	10	10
Arizona Coyotes	24 – 50 – 8	56	2.0	3.3	.0690	.902	1.620	1.557	1.719	3.047	29	30	28	29	29
Boston Bruins	41 – 27 – 14	96	2.5	2.5	.0821	.918	2.480	2.503	2.372	2.373	17	16	19	19	19
Buffalo Sabres	23 – 51 – 8	54	1.9	3.3	.0771	.908	1.544	1.463	1.623	3.099	30	27	30	30	30
Calgary Flames	45 – 30 – 7	97	2.9	2.6	.1052	.911	2.693	2.641	2.708	2.501	8	4	12	13	13
Carolina Hurricanes	30 – 41 – 11	71	2.2	2.7	.0725	.902	1.932	2.172	2.026	2.476	26	28	26	26	26
Chicago Blackhawks	48 – 28 – 6	102	2.7	2.3	.0792	.925	3.072	2.937	2.790	2.165	5	14	3	2	3
Colorado Avalanche	39 – 31 – 12	90	2.5	2.7	.0914	.918	2.454	2.474	2.585	2.629	22	7	21	20	20
Columbus Blue Jackets	42 – 35 – 5	89	2.8	3.0	.0959	.910	2.530	2.371	2.487	2.681	24	10	18	22	22
Dallas Stars	41 – 31 – 10	92	3.1	3.1	.1005	.895	2.549	2.570	2.538	2.480	19	22	15	18	18
Detroit Red Wings	43 – 25 – 14	100	2.8	2.6	.0951	.909	2.579	2.617	2.576	2.411	13	13	13	15	16
Edmonton Oilers	24 – 44 – 14	62	2.4	3.4	.0828	.888	1.613	1.721	1.668	2.814	28	29	29	28	28
Florida Panthers	38 – 29 – 15	91	2.4	2.6	.0787	.912	2.325	2.321	2.430	2.692	23	26	23	24	24
Los Angeles Kings	40 – 27 – 15	95	2.7	2.4	.0860	.911	2.454	2.637	2.473	2.281	12	21	20	14	14
Minnesota Wild	46 – 28 – 8	100	2.8	2.4	.0900	.913	2.767	2.795	2.840	2.419	7	15	9	7	7
Montreal Canadiens	50 – 22 – 10	110	2.6	2.2	.0917	.926	2.905	2.714	2.852	2.519	6	2	6	8	8
Nashville Predators	47 – 25 – 10	104	2.8	2.5	.0865	.913	2.804	2.801	2.726	2.275	10	18	7	6	6
New Jersey Devils	32 – 36 – 14	78	2.1	2.5	.0876	.917	2.039	2.194	2.055	2.485	25	12	25	25	25
New York Islanders	47 – 28 – 7	101	3.0	2.7	.0884	.903	2.758	2.701	2.707	2.405	11	25	10	9	9
New York Rangers	53 – 22 – 7	113	3.0	2.3	.0960	.923	3.074	3.018	2.784	2.036	1	1	2	1	1
Ottawa Senators	43 – 26 – 13	99	2.8	2.5	.0914	.921	2.546	2.694	2.932	2.668	9	6	16	11	11
Philadelphia Flyers	33 – 31 – 18	84	2.6	2.7	.0879	.910	2.081	2.339	2.359	2.594	21	20	24	23	23
Pittsburgh Penguins	43 – 27 – 12	98	2.6	2.5	.0837	.915	2.538	2.600	2.672	2.501	16	19	17	17	15
San Jose Sharks	40 – 33 – 9	89	2.7	2.8	.0865	.907	2.445	2.437	2.441	2.510	20	23	22	21	21
St. Louis Blues	51 – 24 – 7	109	2.9	2.4	.0943	.912	2.971	2.936	2.766	2.107	3	9	4	3	2
Tampa Bay Lightning	50 – 24 – 8	108	3.2	2.5	.1067	.910	2.955	2.922	2.828	2.223	2	3	5	4	4
Toronto Maple Leafs	30 – 44 – 8	68	2.5	3.1	.0859	.906	1.930	2.027	2.486	3.242	27	24	27	27	27
Vancouver Canucks	48 – 29 – 5	101	2.9	2.7	.0962	.910	2.778	2.617	2.473	2.315	15	8	8	16	17
Washington Capitals	45 – 26 – 11	101	2.9	2.4	.0980	.916	2.709	2.851	2.730	2.230	4	5	11	5	5
Winnipeg Jets	43 – 26 – 13	99	2.7	2.5	.0914	.913	2.574	2.675	2.793	2.541	14	11	14	12	12
Home-Ice Advantage							0.188	−0.003	0.022	0.028					

Regression Results

The best fit regression equation for predicting the victory margin as a function of team goals scored and goals allowed is below. The resulting victory margin is positive when the home team is favored and negative when the visiting team is favored.

This model equation is:

$$Est.\ Victory\ Margin = -1.28 + 0.867 \cdot HGS - 0.823 \cdot HGA$$
$$-0.609 \cdot AGS + 0.999 \cdot AGA$$

The signs of the input variables are intuitive. The sign of the sensitivity parameter is expected to be positive for home goals scored (HGS) and away goals allowed (AGA). In both of these cases, the home team will score more goals if these input values are higher. Similarly, the signs of the sensitivity parameters for home goals allowed (HGA) and away team goals scored (AGS) are expected to be negative. The home team is expected to win by fewer goals or possibly lose if the HGA and/or AGS increase in value (Fig. 8.4, Table 8.2).

Performance

The game scores model correctly identified the winner in 787 of 1319 regular-season and playoff games, a rate of 59.7%. Interestingly, accuracy rates for regulation games and overtime games were nearly identical: 61.5% for

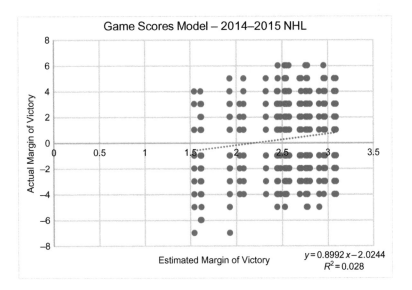

Figure 8.4 Game Scores Model: 2014—15 NHL.

Table 8.2 Game Scores Model: 2014—15 NHL		
Statistic	**Value**	**t-Stat**
b_0	2.685	1.8
b_1 (Home goals scored)	0.653	2.9
b_2 (Home goals allowed)	−1.077	−4.8
b_3 (Visitor goals scored)	−1.287	−5.7
b_4 (Visitor goals allowed)	0.795	3.5
R^2	9.09%	
F-value	32.839	
Standard Error	2.205	

games decided in regulation and 61.3% for games decided in sudden-death overtime. However, it was a different story for games decided by shootout, as the model's favorite won only 47.6% of the time (81 of 170 games).

The model was 79.7% accurate when the win probability was at least 70%, and 70.9% accurate when the win probability was at least 60%. The largest upsets were two wins by the Buffalo Sabres on the road against the Montreal Canadiens, one by shootout in November and the other a 3—2 victory in regulation in February; the model favored the Canadiens by 2.18 goals when hosting Buffalo while giving Montreal an 83.8% win probability.

The closest matchup was the New York Islanders at the Vancouver Canucks, in which the Canucks were favored 50.014% to 49.986% by a margin of 0.00076 goals. Vancouver did win that game in January by a score of 3—2. The widest prediction was for a January game between the Sabres and the New York Rangers at Madison Square Garden. The model put the Rangers' win probability at 86.3% and set the spread at 2.41 goals; the Rangers won that game 6—1.

For the postseason, the model would either favor the same team in each game, or the home team in each game, depending on the difference between the teams' relative strength compared with the home-ice advantage factor. The model was over 50% accurate in eight postseason series, 50% accurate in four series, and less than 50% accurate in three. It picked the winner five times out of six in two first-round series (Calgary Flames vs Vancouver Canucks and Chicago Blackhawks vs Nashville Predators). In the Stanley Cup Finals, the model favored Tampa Bay in every game, 57.1% to 42.9% at home and 51.3% to 48.7% in Chicago. The Blackhawks won the series in six games, as each of the first five games was decided by one goal.

Rankings

The New York Rangers finished atop the game scores model's rankings, ahead of the Eastern Conference Champion Tampa Bay Lightning. The Chicago Blackhawks, who won the Stanley Cup, were ranked No. 5.

The highest-ranked nonplayoff team was the Los Angeles Kings at No. 12, three and six spaces ahead of the Vancouver Canucks and the Anaheim Ducks respectively, two other teams from the Pacific Division that did qualify for the playoffs. Despite winning the Pacific Division, the No. 18 Ducks were the lowest-ranked playoff team. They scored 236 goals and allowed 226 (tied for 11th-most in the NHL), while the Calgary Flames, who won five fewer games than Anaheim, scored five more goals and allowed 10 fewer. The Kings' 205 goals allowed was seventh-best in the NHL, but they also tied for second with 15 overtime losses, and they missed the playoffs by two points in the standings. The No. 17 Boston Bruins were the only other nonplayoff team ranked ahead of a playoff team. Six of the bottom eight represented the Eastern Conference's Atlantic and Metropolitan Divisions.

Example

Our example game to evaluate using the six models will feature two teams from the Eastern Conference, the Pittsburgh Penguins (43−27−12, 98 points) and the Washington Capitals (45−26−11, 99 points), taking place in Washington.

Goal differentials in 2014−15 were 2.64 for and 2.49 against for the Penguins, and 2.89 for and 2.43 against for the Capitals. Based on the betas from the regression, the model predicts a win for the Capitals by more than half a goal, with a win probability of 59.5%:

$$Est.\ Victory\ Margin = 2.685 + 0.653(2.89) - 1.077(2.43)$$
$$- 1.287(2.64) + 0.795(2.49)$$
$$= 0.530 \pm 2.205$$

$$Probability = 1 - NormCDF(0,\ -0.530,\ 2.205) = 59.50\%$$

8.2 TEAM STATISTICS MODEL

In this section, we will demonstrate the regression using two statistics for each team: shooting percentage (goals scored divided by offensive shots on goal) and save percentage (saves divided by opponent shots on goal).

Table 8.3 Team Statistics Model: 2014–15 NHL		
Statistic	**Value**	**t-Stat**
b_0	2.125	0.2
b_1 (Home shooting percentage)	25.600	3.5
b_2 (Home save percentage)	29.782	3.8
b_3 (Away shooting percentage)	−37.163	−5.2
b_4 (Away save percentage)	−30.716	−3.9
R^2	5.62%	
F-value	19.575	
Standard Error	2.247	

The team statistics linear regression model has form:

$$Y = b_0 + b_1 \cdot HTS\% + b_2 \cdot HTSV\% + b_3 \cdot ATS\% + b_4 \cdot ATSV\% + \varepsilon$$

The variables of the model are:

Y = outcome value we are looking to predict (spread, home team goals, away team goals, and total goals)
$HTS\%$ = home team shooting percentage
$HTSV\%$ = home team save percentage
$ATS\%$ = away team shooting percentage
$ATSV\%$ = away team save percentage
b_0, b_1, b_2, b_3, b_4 = model parameter, sensitivity to the variable
ε = model error

The betas of this model are determined from a linear regression analysis as described in Chapter 2, Regression Models. The results of these model are shown in Table 8.3.

Regression Results

The result of the regression model for predicting home team victory margin from our team statistics is:

$$Est. Victory\ Margin = 2.125 + 25.600(HTS\%) + 29.782(HTSV\%)$$
$$- 37.163(ATS\%) - 30.716(ATSV\%) \pm 2.247$$

The model produced an R^2 of 5.6% and a standard error of 2.25 goals. The t-statistics had absolute values between 3.5 and 4, with the exception of the away team's shooting percentage, the t-stat for which was −5.16 (Fig. 8.5).

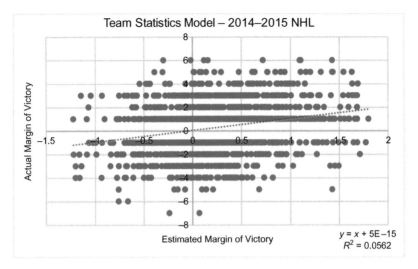

Figure 8.5 Team Statistics Model: 2014–15 NHL.

Performance

The team statistics model's favored teams won 774 of 1319 games, a rate of 58.7%. There were 170 regular-season games decided by shootout, of which the model's favorite won 89 (52.4%). With shootout games ignored, the model was 59.6% accurate. When the win probability was at least 70%, the model was right 77.1% of the time.

The home team was the favorite in 897 games (68.0%). Favored road teams averaged an expected win probability of 56.3%; four (Lightning, Rangers, and Flames twice) had probabilities over 70%, all for games against the Oilers in Edmonton. The biggest favorite was the Rangers, a 78.8% favorite for a home game in February against Arizona, which the Blueshirts did win by a score of 4–3.

Rankings

The Rangers came in as the top-ranked team again, powered by their 9.6% shooting percentage, sixth in the NHL, and .923 save percentage, better than all but the Montreal Canadiens, who were ranked No. 2, and the Chicago Blackhawks. Chicago's 7.92% shooting percentage was fifth worst in the NHL, and as a result the 2015 Stanley Cup champions were ranked No. 14. The only other team in the top 10 in the league in each statistical category was the No. 5. Washington Capitals, fourth in shooting percentage and eighth in save percentage. The Tampa Bay Lightning, champions of the Eastern

Conference, ranked No. 3, followed by the Calgary Flames, the highest-ranked Canadian team. The bottom three places in the rankings were made up by the Carolina Hurricanes, Edmonton Oilers, and Arizona Coyotes.

Example

The Washington Capitals scored on 9.797% of their shots, and their goaltenders stopped 91.6% of their opponents' shots on goal. Pittsburgh's netminders saved goals at a similar rate, 91.5%, but their shooting percentage was much lower at 8.365%.

Note that the betas $b_1 - b_4$ are based on the shooting and save percentages expressed as decimals between 0 and 1, and not as percentages between 0 and 100. So we need to provide the Penguins' shooting percentage as 0.08365, and we find that Washington should win this game by 0.7 goals:

$$Est.\ Margin\ of\ Victory = 2.125 + 25.600(0.09797) + 29.782(0.91596)$$
$$- 37.163(0.08365) - 30.716(0.91504)$$
$$\pm SE = 0.69739 \pm 2.247$$

The margin of -0.69739 goals is equivalent to a win probability of 62.1% for the Capitals:

$$p = 1 - NormCDF(0,\ 0.69739,\ 2.247) = 62.1\%$$

8.3 LOGISTIC PROBABILITY MODEL

The logistic model is as follows:

$$y^* = \frac{1}{1 + \exp\{-(b_0 + b_h - b_a)\}}$$

Here, b_0 denotes a home-field advantage parameter, b_h denotes the home team rating parameter value, and b_a denotes the away team rating parameter value. The value y^* denotes the probability that the home team will win the game. Team ratings for the logistic probability are determined via maximum likelihood estimates and are shown Table 8.1.

Estimating Spread

The estimated spread (i.e., home team victory margin) is determined via a second analysis where we regress the actual home team spread on the estimated probability y^* (as the input variable). This regression has form:

$$Actual\ Spread = a_0 + a_1 \cdot y^*$$

This model now provides a relationship between the logistic home team winning probability and the home team winning percentage. It is important to note here that analysts may need to incorporate an adjustment to the spread calculation if the data results are skewed (see Chapter 3: Probability Models).

The solution to this model is (Table 8.4):

Table 8.4 Logistic Probability Regression: 2014−15 NHL		
Statistic	Value	t-Stat
a_0	−2.331	−9.4
a_1	4.733	10.7
b_0 (Home-ice advantage)	0.188	
R^2	8.05%	
F-value	115.335	
Standard Error	2.215	

Therefore, after computing the home team winning probability, the expected spread is estimated from the following equation based on the regression results:

$$Estimated\ Spread = -2.331 + 4.733 \cdot Probability$$

A graph illustrating the estimated spread from the probability estimates is shown in Fig. 8.6.

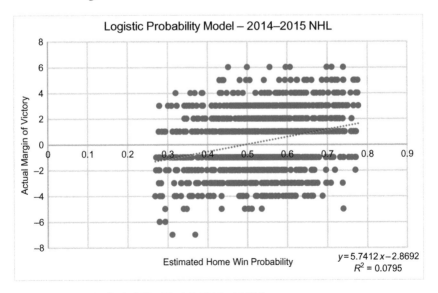

Figure 8.6 Logistic Probability Model: 2014−15 NHL.

Performance

The logistic probability model's accuracy rate for predicting winners was 60.9%, its favorites winning 803 of 1319 regular-season and playoff games. The winning percentage for games not decided by shootout was 61.8% (710 out of 1149). The home teams were favored in 65% of games, of which they won 61.5%. Visiting teams were favored 35% of the time, and won 59.7%. A projected spread of one goal was equivalent to a win probability of 67.4%.

The model's most even matchup was for the Minnesota Wild playing at the Dallas Stars. Dallas was favored by one one-hundred-thousandth of a goal, and their win probability was 50.0001%. The teams met in Dallas three times during the 2014−15 season; the first game went to the Wild 2−1, while the Stars took the second 5−4 in overtime before blowing out Minnesota 7−1 in the rubber game. The 36 most lopsided matchups involved various teams hosting the Buffalo Sabres, Edmonton Oilers, and Arizona Cardinals; the favorites won 28 of the 36 games while splitting six shootouts.

Rankings

Only one nonplayoff team finished ahead of a playoff team; the Dallas Stars' rating of 2.579 put them at No. 15, behind all eight playoff-bound Western Conference teams but ahead of Ottawa and Pittsburgh, the Eastern Conference's seven- and eight-seeds. Anaheim took the top spot in the rankings with a team rating of 3.092, ahead of the New York Rangers (3.074) and the eventual Stanley Cup Champion Chicago Blackhawks (3.072). The Tampa Bay Lightning, Chicago's Finals opponent, was ranked No. 5. The eight teams with the lowest point totals in the standings finished in the same order in the model's rankings.

Example

The Capitals' logistic rating of 2.709 was 12th in the NHL, ranking them between the New York Islanders and Calgary Flames. Pittsburgh was ranked 18th with a logistic rating of 2.538.

$$Probability\ Parameter = \frac{1}{1 + e^{-(0.188 + 2.709 - 2.538)}} = \frac{1}{1 + e^{0.359}} = 0.5889$$

We can then take this value and combine it with the regression parameters b_0 (−2.331) and b_1 (4.733) to estimate that this matchup should result in a Washington victory by a little under half a goal, with win probabilities of 58.2% for the Capitals and 41.8% for the Penguins:

$$Estimated\ Spread = -14.003 + 28.466 \cdot 0.5889 \pm SE = -0.456 \pm 2.215$$

$$Probability = 1 - normdist(0,\ -0.456,\ 2.215) = 58.16\%$$

8.4 TEAM RATINGS MODEL

The team ratings prediction model is a linear regression model that uses the team ratings determined from the logistic probability model as the explanatory variables to estimate home team victory margin. This is one of the reasons why the logistic model is one of the more important sports models since its results can be used in different modeling applications.

The team ratings regression has form:

$$Y = b_0 + b_1 \cdot Home\ Rating + b_2 \cdot Away\ Rating + \epsilon$$

The variables of the model are:

$Home\ Rating$ = home team strength rating (from logistic probability model)

$Away\ Rating$ = away team strength rating (from logistic probability model)

b_0 = home-field advantage value

b_1 = home team rating parameter

b_2 = away team rating parameter

Y = home team's victory margin (positive indicates home team is favored and negative value indicates away team is favored)

The probability of winning is determined from:

$$Probability = 1 - NormCDF(0, Y, SE)$$

The betas of this model, b_1 and b_2, are determined from a linear regression analysis as described in Chapter 2, Regression Models.

Regression Results

The best first regression equation to predict home team victory margin from team strength ratings is:

$$Victory\ Margin = 1.1174 + 0.9269 \cdot Home\ Rating - 1.2738 \cdot Away\ Rating$$

Table 8.5 Team Ratings Model: 2014–15 NHL		
Statistic	Value	t-Stat
b_0	1.117	2.2
b_1 (Home team rating)	0.927	6.5
b_2 (Away team rating)	−1.274	−8.9
R^2	8.47%	
F-value	60.484	
Standard Error	2.215	

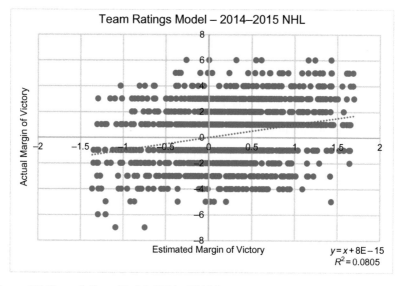

Figure 8.7 Team Ratings Model: 2014–15 NHL.

The regression results and a graph showing the actual victory margin as a function of estimated victory margin are given in Table 8.5 and Fig. 8.7, respectively.

Performance

The ratings regression chose the winner in 800 of 1311 games, or 61.0% of the time. An estimated spread of one goal was equivalent to a win probability of 67.45%.

The 56 highest win probabilities were projected for various teams playing on their home ice against either the Sabres, Coyotes, or Ducks. That figure rises to 95 if the Maple Leafs and Hurricanes are included, ranging from 73.6% (Hurricanes at Lightning) to 81.9% (Sabres at

Ducks). Over those 95 games, the favorites were a combined 77–11–7. The lowest win probability for a favorite was 50.1% for the New York Islanders on the road for two games against the Detroit Red Wings in December and January, each team winning one game.

The biggest upset was a 3–2 shootout win by the Coyotes against the Ducks in Anaheim, with Arizona given only 19.2% probability of winning. On November 9, the Oilers beat the Rangers 3–1 at Madison Square Garden; the model had given New York an 80.7% probability of victory, projecting a spread of 1.91 goals.

Example

The ratings regression also predicts a narrow win by the Capitals at home over Pittsburgh:

$$Est.\ Victory\ Margin = -1.117 + 0.927(2.709) - 1.274(2.538)$$
$$\pm SE = -0.395 \pm 1.274$$

The estimated winning probability is 42.9% for the Penguins and 57.1% for Washington:

$$Probability = 1 - NormCDF(0, -0.395, 1.274) = 57.1\%$$

8.5 LOGIT SPREAD MODEL

The logit spread model is a probability model that predicts the home team victory margin based on an inferred team rating metric. The model transforms the home team victory margin to a probability value between 0 and 1 via the cumulative distribution function and then estimates model parameters via logit regression analysis.

The logit spread model has following form:

$$y = b_0 + b_h - b_a$$

where b_0 denotes a home-field advantage parameter, b_h denotes the home team parameter value, and b_a denotes the away team parameter value.

The left-hand side of the equation y is the log ratio of the cumulative density function of victory margin (see chapter: Sports Prediction Models, for calculation process).

In this formulation, the parameters values b_0, b_h, b_a are then determined via ordinary least squares regression analysis. The results of this analysis are shown in Table 8.1.

Estimating Spreads

Estimating the home team winning margin is accomplished as follows.

If team k is the home team and team j is the away team, we compute y using the logit parameters:

$$y = b_0 + b_k - b_j$$

Compute y^* from y via the following adjustment:

$$y^* = \frac{e^y}{1 + e^y}$$

Compute z as follows:

$$z = norminv(y^*)$$

And finally, the estimated home team spread:

$$Estimated\ Spread = \bar{s} + z \cdot \sigma_s$$

where

$\bar{s}$ = average home team winning margin (spread) across all games

σ_s = standard deviation of winning margin across all games

Estimating Probability

The corresponding probability of the home team winning is determined by performing a regression analysis of actual spread as a function of estimated spread to determine a second set of model parameters. This model has form:

$$Actual\ Spread = a_0 + a_1 \cdot Estimated\ Spread$$

To run this regression, we need to compute the estimated spread for all games using the logit spread parameters from above (see Table 8.1).

The solution to this model is (Table 8.6):

Table 8.6 Logit Spread Model: 2014–15 NHL	
Statistic	**Value**
$\bar{s}$ (Average home victory margin)	0.242
σ_s (Home victory margin standard deviation)	2.309
b_0 (Home-ice advantage)	− 0.003
R^2	8.99%
F-value	130.145
Standard Error	2.204

A graphical illustration of this model is (Fig. 8.8):

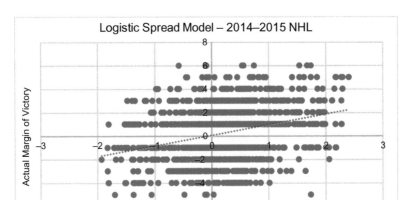

Figure 8.8 Logit Spread Model: 2014—15 NHL.

Performance

The logit spread model was able to predict the winners 797 times in 1319 games, a rate of 60.4%. The winning percentage was 62.3% (716 of 1149 games) excluding games decided by shootout. Favorites with win probabilities over 70% won 166 of 212 games, a rate of 78.3%. A spread of one goal equaled a win probability of 67.5%. The most lopsided matchup was the Sabres playing the Rangers in Madison Square Garden; New York was favored by 2.4 goals with a win probability of 86.2%, and went on to defeat Buffalo by a score of 6—1. The biggest upset was a 2—1 Buffalo victory in Washington in November; the model gave the Sabres only a 16.1% chance. The Capitals would avenge the loss in a March rematch, also by a 6—1 score.

Rankings

At No. 14, the Los Angeles Kings were the only nonplayoff team to finish ahead of a playoff team; their logit spread rating of 2.637 was higher than that of the Detroit Red Wings (2.6168, No. 6 seed in the East), Vancouver Canucks (2.6166, No. 5 seed in the West), and the Pittsburgh Penguins (2.600, No. 8 seed in the East). The New York Rangers took the top spot in the rankings, followed by the eventual Stanley Cup Champion Chicago Blackhawks, the fourth seed in the Western

Conference. The St. Louis Blues, the West's 1-seed, were ranked third while the Anaheim Ducks, the 2-seed in the West, ranked tenth.

Example

Washington had a logit spread rating of 2.851, fifth best in the NHL. Pittsburgh, on the other hand, had the lowest rating of any playoff team, as their 2.600 was 17th. The home-ice advantage factor for the logit spread model (b_0) was a negligible -0.003 for the NHL in 2014−15.

Home teams averaged a goal differential of $+0.242$ with a standard deviation of 2.309. With those parameters, the logit spread model provides a 0.6-goal advantage for Washington, with a 60.7% chance of defeating the Penguins at home.

$$y^* = \frac{1}{1 + e^{-(2.851 - 2.600 - 0.003)}} = \frac{1}{1 + e^{-0.248}} = 0.5616$$

$$Victory\ Margin = z \cdot \sigma_s + \bar{s} = norminv(0.5616) \cdot 2.309 + 0.242 = 0.5998$$

$$Probability = 1 - normdist(0, 0.5998, 2.204) = 60.73\%$$

8.6 LOGIT POINTS MODEL

The logit points model is a probability model that predicts the home team victory margin by taking the difference between home team predicted goals and away team predicted goals. The predicted goals are determined based on inferred team "ratings" similar to the logit spread model discussed above.

The logit points model has following form:

$$h = c_0 + c_h - c_a$$

$$a = d_0 + d_h - d_a$$

where

h is the transformed home team goals, c_0 denotes a home-field advantage, c_h denotes the home team rating, and c_a denotes the away team rating corresponding to home team goals.

a is the transformed away team goals, d_0 denotes a home-field advantage, d_h denotes the home team rating, d_a denotes the away team rating corresponding to away team goals.

The left-hand side of the equation h and a is the log ratio of the cumulative density function of home team goals and away team goals respectively (see chapter: Sports Prediction Models, for a description).

Estimating Home and Away Team Goals

Estimating the home team goals is accomplished directly from the home goals team ratings. These rating parameters are shown in Table 8.1. If team k is the home team and team j is the away team, the transformed home team goals are:

$$h = c_0 + c_k - c_j$$

$$h^* = \frac{e^h}{1 + e^h}$$

Then

$$z = norminv(h^*)$$

And finally, the x-value is:

$$Home\ Points = \overline{h} + z \cdot \sigma_h$$

where

$\overline{h} =$ average home team goals
$\sigma_h =$ standard deviation of home team goals

Away goals are estimated in the same manner but using the team ratings for the away goals model.

Estimating Spread

The estimated home team victory margin is computed directly from the home team goals and away team goals as follows:

$$Est.\ Spread = Home\ Team\ Goals - Away\ Team\ Goals$$

Estimating Probability of Winning

The corresponding probability of winning is determined by performing a regression analysis of actual spread as a function of estimated spread. The model has form:

$$Actual\ Spread = b_0 + b_1 \cdot Est.\ Spread$$

The solution to this model is (Table 8.7):

Table 8.7 Logit Points Model: 2014–15 NHL		
Statistic	Home	Away
$\bar{s}$ (Average score)	2.838	2.596
σ_s (Score standard deviation)	1.655	1.526
b_0 (Home-ice advantage)	0.022	0.028
R^2		9.00%
F-value		130.283
Standard Error		2.204

A graphical illustration of this model is (Fig. 8.9):

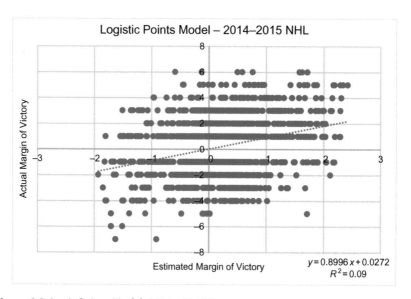

Figure 8.9 Logit Points Model: 2014–15 NHL.

Performance

The logit points model predicted the winner in 796 games, a rate of 60.3%. Excluding games decided by shootout, the rate was 62.2% (715 of 1149). The Sabres, Coyotes, Maple Leafs, and Oilers comprised the 102 least-favored home teams (from 19.0% for the Sabres hosting the Rangers to 33.1% for the Maple Leafs hosting the Capitals) as well as the 96 least-favored visitors (from 13.7% for Buffalo at the Rangers to 25.7% for Toronto at Washington); the four teams won only 24.2% of those 198 games.

The predicted spread was within ± 1 goal of the actual margin of victory in 402 games (31.5%). The predicted spread was one goal or less in 1048 games (79.5%), with a one-goal favorite having a win probability of 67.4%.

Example

The ratings for the Capitals at home were $b_h = 2.7302$ and $d_h = 2.2305$, and the Penguins' were $b_a = 2.2674$ and $d_a = 2.5007$. The home-ice advantage had factors of $b_0 = 0.0220$ and $d_0 = 0.0281$. The model's conclusion is that the Capitals should win by a little over half a goal, and that the probability that Washington would defeat Pittsburgh at home is 60%.

$$Home\ Score = norminv\left(\frac{1}{1 + e^{-(0.0220 + 2.7302 - 2.2674)}}\right) \cdot 1.6553 + 2.8378 = 2.9205$$

$$Away\ Score = norminv\left(\frac{1}{1 + e^{-(0.0281 + 2.2305 - 2.5007)}}\right) \cdot 1.5261 + 2.5959 = 2.3645$$

$$Victory\ Margin = 2.9205 - 2.3645 = 0.5559$$

$$Probability = 1 - normcdf(0, 0.5559, 2.204) = 59.96\%$$

8.7 EXAMPLE

In our matchup of the Capitals and Penguins in Washington, all six models favor the Capitals. The six win probabilities were all very similar, fitting in a narrow five-percentage-point range of 57.1−62.1%. Four of the six models favor Washington by more than half a goal. The game scores model was closest to the average of the six both for win probability and margin of victory (Table 8.8).

Table 8.8 Example Results

Model	Favorite	Underdog	Line	P(WAS Win)	P(PIT Win)
Game Scores	Washington	Pittsburgh	0.5	59.5%	40.5%
Team Statistics	Washington	Pittsburgh	0.7	62.2%	37.8%
Logistic Probability	Washington	Pittsburgh	0.5	58.2%	41.8%
Team Ratings	Washington	Pittsburgh	0.4	57.1%	42.9%
Logit Spread	Washington	Pittsburgh	0.6	60.7%	39.3%
Logit Points	Washington	Pittsburgh	0.6	60.0%	40.0%
Average	Washington	Pittsburgh	0.5	59.6%	40.4%

The Capitals hosted the Penguins twice in the 2014−15 regular season. Washington shut Pittsburgh out on January 28 by a score of 4−0, a loss the Penguins avenged with a 4−3 win on February 25.

8.8 OUT-SAMPLE RESULTS

We performed an out-sample analysis where we predicted our game results using a walk forward approach. In this analysis, we use previous game results data to predict future games. Here, the model parameters were estimated after about 10 games per team and then we predicted the winning team for the next game. For all models we found the predictive power declined slightly using an out-sample test, but after about 20 games per team the results from the out-sample began to converge to the results with in-sample data.

These models had a decrease in winning percentage using out-sample results of only −0.5%. The winning percentage for these models after about 20 games and into the postseason were in line with the in-sample results, thus indicating that once we have about 20 games per team the model results are as strong as they can be (Fig. 8.10).

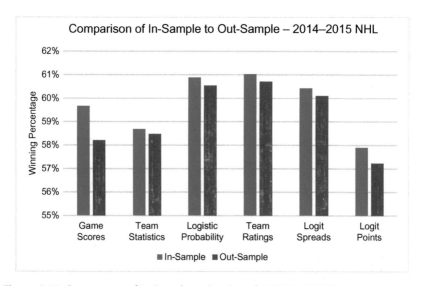

Figure 8.10 Comparison of In-Sample to Out-Sample: 2014−15 NHL.

8.9 CONCLUSION

In this chapter we applied six different sports model approaches to National Hockey League results for the 2014–15 season. The models were used to predict winning team, estimated home team victory margin, and probability of winning. Overall the six models performed fairly similarly to one another; each had in-sample winning team prediction rates between 58% and 61% and regression errors between 2.20 and 2.25. As would be expected, each model's predictive accuracy was diminished by games decided by shootout. Five of the six models had similar regression R^2 values of 8.1–9.1%, while the team statistics model, using shooting and save percentages as parameters, was lower at 5.6%. The three most accurate models were all probability models (logistic probability, team ratings, and logit spreads), followed by the two data-driven models (game scores and team statistics) and the fourth probability model.

CHAPTER 9

Soccer—MLS

In this chapter, we apply the sports modeling techniques introduced in previous chapters to the 2015 Major League Soccer (MLS) season. Our goal is to provide readers with proper techniques to predict expected winning team and probability of winning, and to estimate winning victory margin.

Our techniques include six different models: game scores, team statistics, team ratings, logistic probability, logit spreads, and logit points. These models are based on linear regression techniques described in Chapter 2, Regression Models, and on probability estimation methods described in Chapter 3, Probability Models.

The models were evaluated in three different ways based on in-sample data: winning percentage, R^2 goodness of the estimated victory margin, and the regression error. Out-sample performance results are discussed at the end of the chapter.

Fig. 9.1 depicts the winning percentage by model. For the purposes of comparing the various models, we will calculate the favorites' winning percentage by counting a draw as half a win and half a loss. This is because tie games are so common in Major League Soccer; in 2015 more than 20% of matches ended in draws. All of the six models were between 64.0% and 66.1% accurate in correctly identifying the winner, compared with 62.7% for the Vegas line. The low-scoring nature of soccer matches affected the other two performance analyses. Fig. 9.2 illustrates the R^2 goodness of fit for each model, which was uniformly low, between 2.7% and 5.0%. Fig. 9.3 depicts the regression error surrounding the predicted victory margin. All of the models' regression errors were within 0.03 goals of each other. The Vegas line does not set point spreads for soccer and thus they are not included in Figs. 9.2 and 9.3 (Table 9.1).

9.1 GAME SCORES MODEL

The game scores regression model predicts game outcomes based on the average number of goals scored and allowed by each team.

Optimal Sports Math, Statistics, and Fantasy.
DOI: http://dx.doi.org/10.1016/B978-0-12-805163-4.00009-8

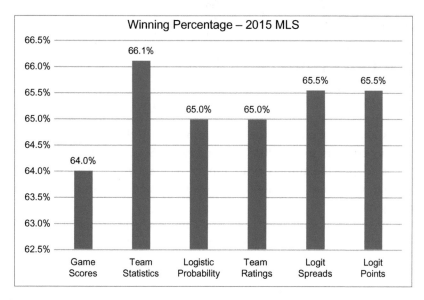

Figure 9.1 Winning Percentage: 2015 MLS.

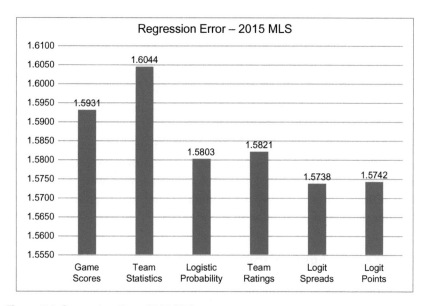

Figure 9.2 Regression Error: 2015 MLS.

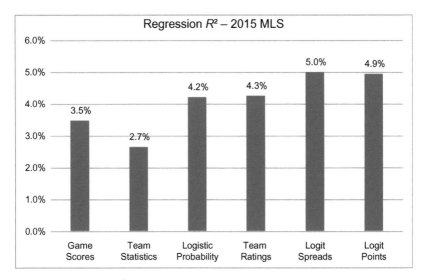

Figure 9.3 Regression R^2: 2015 MLS.

The game scores model has form:

$$Y = b_0 + b_1 \cdot HGS + b_2 \cdot HGA + b_3 \cdot AGS + b_4 \cdot AGA + \varepsilon$$

In the representation, the dependent variable Y denotes the value that we are trying to predict; this can be the home team's margin of victory (or defeat), the number of goals scored by the home team, the number of goals scored by the away team, and the total combined score. For our purposes, we define the Y variable to be the home team victory margin. A positive value indicates the home team won by the stated number of goals and a negative value indicates the home team lost by the stated number of goals.

The variables of the model are:

$Y =$ home team victory margin
$HGS =$ home team average goals scored per game
$HGA =$ home team average goals allowed per game
$AGS =$ away team average goals scored per game
$AGA =$ away team average goals allowed per game
$b_0, b_1, b_2, b_3, b_4 =$ model parameters, represents the sensitivities to the corresponding model factor

The betas of this model are determined from a linear regression analysis as described in Chapter 2, Regression Models.

Table 9.1 Regression Results and Rankings

Team	Team Statistics						Ratings				Rankings				
	W – L – T	PF/G	PA/G	SOG/G	CK/G	OFF/G	Logistic Rating	Logit Spread	Logit Home Points	Logit Away Points	Game Scores	Team Statistics	Logistic Ratings	Logit Spread	Logit Points
Chicago Fire	8 – 20 – 6	1.26	1.56	4.79	5.79	2.38	1.754	2.077	1.924	2.385	15	8	20	19	19
Colorado Rapids	9 – 15 – 10	0.97	1.12	3.74	5.85	2.79	2.084	2.323	2.579	2.877	20	20	17	15	15
Columbus Crew	15 – 11 – 8	1.71	1.53	5.29	6.09	2.06	2.633	2.610	2.833	2.778	3	1	6	9	9
DC United	15 – 13 – 6	1.26	1.94	4.00	4.79	2.62	2.444	2.362	2.844	3.063	18	16	13	14	14
FC Dallas	18 – 10 – 6	1.53	1.21	4.35	4.94	1.68	2.993	2.881	2.854	2.374	5	11	3	2	2
Houston Dynamo	11 – 14 – 9	1.24	1.21	3.65	4.76	1.44	2.368	2.364	2.662	2.953	13	17	15	13	13
Los Angeles Galaxy	14 – 11 – 9	1.65	1.35	4.50	4.68	1.41	2.877	2.881	2.901	2.470	4	9	4	3	3
Montreal Impact	15 – 13 – 6	1.41	1.15	4.97	4.82	1.94	2.551	2.616	2.710	2.572	7	4	10	8	8
New England Revolution	14 – 12 – 8	1.41	1.47	4.65	5.82	1.85	2.594	2.416	2.047	2.073	9	10	8	11	11
New York City	10 – 17 – 7	1.44	1.24	4.71	4.94	2.29	1.901	2.152	2.706	3.297	6	7	19	17	17
New York Red Bulls	18 – 10 – 6	1.82	1.24	5.24	6.06	3.26	3.130	2.991	2.865	2.298	1	3	1	1	1
Orlando City	12 – 14 – 8	1.35	1.53	3.68	4.82	2.41	2.380	2.103	2.008	2.362	11	18	14	18	18
Philadelphia Union	10 – 17 – 7	1.24	1.24	4.38	5.35	2.06	2.040	2.075	2.028	2.551	14	13	18	20	20
Portland Timbers	15 – 11 – 8	1.21	1.41	4.88	6.00	2.29	3.108	2.703	2.479	2.113	17	6	2	6	6
Real Salt Lake	11 – 15 – 8	1.12	1.68	4.24	4.53	2.06	2.467	2.208	2.074	2.446	19	14	12	16	16
San Jose Earthquakes	13 – 13 – 8	1.21	1.38	3.79	5.76	2.18	2.590	2.630	2.056	1.813	16	19	9	7	7
Seattle Sounders	15 – 13 – 6	1.29	1.41	4.24	4.26	1.97	2.676	2.833	2.719	2.332	12	12	5	4	4
Sporting Kansas City	14 – 11 – 9	1.41	1.41	4.15	4.50	1.74	2.599	2.557	2.525	2.382	8	15	7	10	10
Toronto FC	15 – 15 – 4	1.71	1.29	4.97	5.26	1.97	2.310	2.401	2.223	2.215	2	5	16	12	12
Vancouver Whitecaps	16 – 13 – 5	1.32	1.12	5.06	5.09	2.06	2.500	2.795	2.883	2.560	10	2	11	5	5
Home-Field Advantage							0.726	0.024	0.078	0.087					

This model can also be used to calculate the probability p that the home team will win the game as follows:

$$p = 1 - NormCDF(0, Y, SeY)$$

Here, Y and SeY are the expected victory margin and regression error term respectively, and zero indicates the reference point used for the calculation (i.e., the probability that the winning margin will be greater than zero).

Regression Results

The best fit regression equation for predicting the victory margin as a function of team goals scored and goals allowed is below. The resulting victory margin is positive when the home team is favored and negative when the visiting team is favored.

This model equation is:

$$Est.\ Victory\ Margin = -2.213 + 1.210 \cdot HGS + 0.345 \cdot HGA$$
$$-0.208 \cdot AGS + 0.696 \cdot AGA$$

The R^2 value of the game scores model's margin of victory model was low at 3.48%. The only significant t-Stat was 3.012 for home team goals allowed.

The signs of the input variables are intuitive. The sign of the sensitivity parameter is expected to be positive for home goals scored (HGS) and away goals allowed (AGA). In both of these cases, the home team will score more goals if these input values are higher. Similarly, the signs of the sensitivity parameters for home goals allowed (HGA) and away team goals scored (AGS) are expected to be negative. The home team is expected to win by fewer goals or possibly lose if the HGA and/or AGS increase in value.

Table 9.2 shows the results of the regression model and Fig. 9.4 shows a graph of actual victory margin (spreads) compared to the model's estimates.

Performance

The game scores regression's favorites combined for a record of 194−89−74. It correctly predicted the winner 54.3% of the time; with the draws counted as half a win and half a loss, the rate was 64.0%. The Vegas favorite won 187 matches and lost 96, a rate of 62.7% counting the 74 draws as each half a win and half a loss. The game scores regression predicted the winner 194 times (64.7%), an improvement of seven matches.

Table 9.2 Game Scores Model: 2015 MLS

Statistic	Value	t-Stat
b_0	−2.214	−1.9
b_1 (Home goals scored)	1.210	3.0
b_2 (Home goals allowed)	0.345	0.8
b_3 (Visitor goals scored)	−0.207	−0.5
b_4 (Visitor goals allowed)	0.696	1.6
R^2	3.48%	
F-value	3.170	
Standard Error	1.593	

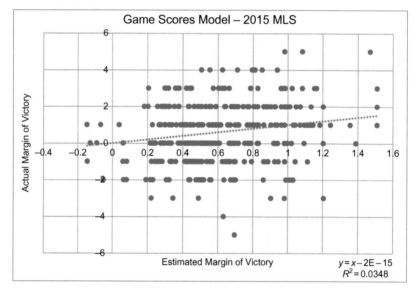

Figure 9.4 Game Scores Model: 2015 MLS.

If the model is instructed to predict a draw when the projected margin of victory is less than half a goal, and only predict a winner when the margin is at least half a goal, the favorite won 59% of the matches, lost 21%, and played to a draw 20% of the time. As the probability of victory is directly related to the estimated margin of victory, a spread of half a goal is equivalent to a win probability of 62.3%.

$$p = 1 - NormCDF(0, 0.5, 1.593) = 62.3\%$$

The home-pitch advantage was substantial enough that the visiting team was favored only eight times in 357 matches (2.24%). Colorado was the home team in all eight matches in which the road team was favored; the Rapids combined for a record of 2—3—3 in those eight matches. Overall the Rapids had the 14 lowest win probabilities for home teams.

The New York Red Bulls were the beneficiaries of the highest calculated probability of victory, 82.8%, for each of three matches with DC United that they hosted, all of which they won. There were 17 matches in which the favorite's probability was at least 76%; the favorites won 14 and lost 1.

During the playoffs, the model's favorites collectively recorded 12 wins, 3 losses, and 2 draws (70.6%).

The range for the estimated victory margins spanned 0.04 goals (Colorado Rapids a narrow favorite at home over Sporting Kansas City) to just over 1.5 (New York Red Bulls hosting DC United). Actual margins of victory went as high as five goals, though 65.5% of matches were decided by only one or two goals and 20.7% ended in draws.

The estimated goal spreads were accurate to within half a goal in 90 matches (25.2%), one goal in 172 matches (48.2%), and two goals in 285 matches (79.8%). The projections missed by more than three goals 23 times (6.4%); the largest discrepancy was a 5—0 win by the San Jose Earthquakes over Sporting Kansas City, a match the model had called for Kansas City by 0.7 goals.

Rankings

Twelve teams qualify for the playoffs in the 20-team league. In the game scores regression, there were 10 playoff teams among the top 12. The two playoff teams that were ranked lower than No. 12 were the Portland Timbers at No. 17 and DC United at No. 18. The regression weighted offense (goals scored per game) much more heavily than defense (goals allowed per game); Portland averaged 1.21 goals per game, tied for third lowest in MLS. DC United was tied for seventh-lowest at 1.26, while they were last in MLS with 1.94 goals allowed per game. New York City was sixth in offense and tied for sixth in defense, but did not make the playoffs by virtue of their 10—17—7 record; they won by an average 1.6 goals and lost by an average of 1.4.

Example

The example matchup for MLS will pit the New York Red Bulls (4—12) against the Toronto Football Club (FC) on Toronto's home pitch. The Red Bulls led MLS by scoring 1.87 goals per game in 2015; Toronto FC was tied for second with 1.71. Defensively, the Red Bulls' 1.24 goals allowed per game put them in a three-way tie with New York City and Philadelphia Union for the sixth-best defense in the league; Toronto was right behind in ninth place with 1.29. The game scores regression projects that Toronto would win by 0.78 goals, plus or minus the standard error of 1.593 goals, with a win probability of 68.8%:

$$Victory\ Margin = -2.213 + 1.21 \cdot 1.71 + 0.345 \cdot 1.30 - 0.208 \cdot 1.83 \\ + 0.696 \cdot 1.24 \pm SE = 0.779 \pm 1.593$$

$$p = 1 - NormCDF(0,\ 0.779,\ 1.593) = 68.8\%$$

9.2 TEAM STATISTICS MODEL

For soccer, we will provide three per-game rates to the team statistics model: shots on goal, corner kicks, and offsides penalties.

The team statistics linear regression model has form:

$$Y = b_0 + b_1 \cdot HTSPG + b_2 \cdot HTCKPG + b_3 \cdot HTOFFPG + b_4 \cdot ATSPG \\ + b_5 \cdot ATCKPG + b_6 \cdot ATOFFPG + \varepsilon$$

The variables of the model are:

Y = outcome value we are looking to predict
$HTSPG$ = home team shots on goal per game (offense)
$HTCKPG$ = home team corner kicks per game
$HTOFFPG$ = home team offsides penalties per game
$ATSPG$ = away team shots on goal per game (offense)
$ATCKPG$ = away team corner kicks per game
$ATOFFPG$ = away team offsides penalties per game
$b_0, b_1, b_2, b_3, b_4, b_5, b_6$ = model parameter, sensitivity to the variable
ε = model error

The betas of this model are determined from a linear regression analysis as described in Chapter 2, Regression Models. The results of these model are shown in Table 9.3.

Table 9.3 Team Statistics Model: 2015 MLS		
Statistic	**Value**	**t-Stat**
b_0	1.425	1.1
b_1 (Home shots on goal per game)	0.387	2.1
b_2 (Home corner kicks per game)	−0.126	−0.7
b_3 (Home offsides per game)	−0.272	−1.2
b_4 (Away shots on goal per game)	−0.116	−0.6
b_5 (Away corner kicks per game)	−0.052	−0.3
b_6 (Away offsides per game)	−0.249	−1.1
R^2	2.65%	
F-value	−1.090	
Standard Error	1.604	

Regression Results

The result of the regression model for predicting home team victory margin from our team statistics is:

$$Victory\ Margin = 1.425 + 0.387 \cdot HTSPG - 0.126 \cdot HTCKPG$$
$$- 0.272 \cdot HTOFFPG - 0.116 \cdot ATSPG$$
$$- 0.052 \cdot ATCKPG - 0.249 \cdot ATOFFPG \pm 1.604$$

The victory margin prediction model produced an R^2 of 2.65% and a standard error of 1.60 goals. The one average with a significant t-statistic was the home team's shots on goals per game.

These regression results are listed in Table 9.3, with a graph of actual spread compared to predicted spread in Fig. 9.5.

Performance

The team statistics regression's favorites won 199 matches and lost 84, a success rate of 66.1%. The Vegas favorite won 187 matches and lost 96, an adjusted rate of 62.7% excluding 74 draws.

With six, this regression had even fewer underdog home teams than the eight returned by the game scores regression. This includes a matchup of New York City at Colorado, where the visiting New York squad's win probability was calculated to be just 50.018%. The lowest win probability for a home team was Orlando, given a 46.3% chance of beating the Red Bulls at home. The highest was 76.6% for Los Angeles the two times they hosted Houston, a 1−1 draw in March and a 1−0 Galaxy victory in May.

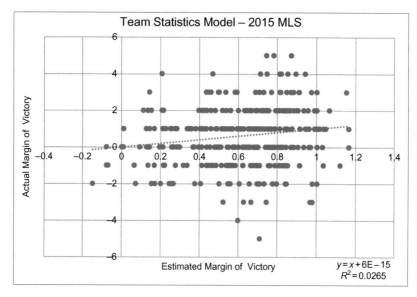

Figure 9.5 Team Statistics Model: 2015 MLS.

Estimated victory margins ranged from 0.001 goals (New York City at Colorado Rapids) to 1.163 goals (Los Angeles at home against Houston).

The estimated goal spreads were accurate to within half a goal in 90 matches (25.2%), one goal in 172 matches (48.2%), and two goals in 285 matches (79.8%). The projections missed by more than three goals 23 times (6.4%); the largest discrepancy was a 5−0 win by the San Jose Earthquakes over Sporting Kansas City, a match the model had called for Kansas City by 0.7 goals.

Rankings

The top six spots in the rankings went to playoff teams, as did 10 of the top 11. No. 7 New York City was the one nonplayoff team in the Top 10, primarily because their 4.71 shots on goal average, the stat with the most significance according to the regression, was good enough for eighth in MLS. The Red Bulls' shot average of 5.24 and 6.06 corner kicks per game were each second only to Columbus's, but they were held back by virtue of their league-worst 3.27 offsides penalty average. Despite their 15 wins, DC United was 16th in shooting, 15th in corner kicks, and 18th in offsides penalties.

Example

The Red Bulls were second in MLS with 5.24 shots on goal and 6.06 corner kicks per game, while Toronto FC was fourth with 4.97 shots on goal and ninth with 5.26 corner kicks per match. The Red Bulls led the league by a comfortable margin with 3.26 offsides penalties per match, nearly half a penalty per game more than any other team, and far more than Toronto's 1.97. Thus under the team statistics model, Toronto is favored by less than half a goal:

$$
\begin{aligned}
\textit{Est. Margin of Victory} = {}& 1.425 + 0.387 \cdot 4.97 - 0.126 \cdot 5.26 \\
& - 0.272 \cdot 1.97 - 0.116 \cdot 5.24 - 0.052 \cdot 6.06 \\
& - 0.249 \cdot 3.26 \pm SE = 0.414 \pm 1.604
\end{aligned}
$$

The margin of 0.414 goals is equivalent to a win probability of 60.2% for Toronto FC:

$$
p = 1 - NormCDF(0,\ 0.414,\ 1.604) = 60.2\%
$$

9.3 LOGISTIC PROBABILITY MODEL

The logistic model is as follows:

$$
\gamma^* = \frac{1}{1 + \exp\big\{-(b_0 + b_h - b_a)\big\}}
$$

Here, b_0 denotes a home-field advantage parameter, b_h denotes the home team rating parameter value, and b_a denotes the away team rating parameter value. The value γ^* denotes the probability that the home team will win the game. Team ratings for the logistic probability are determined via maximum likelihood estimates and are shown in Table 9.1.

Estimating Spread

The estimated spread (i.e., home team victory margin) is determined via a second analysis where we regress the actual home team spread on the estimated probability γ^* (as the input variable). This regression has form:

$$
\textit{Actual Spread} = a_0 + a_1 \cdot \gamma^*
$$

This model now provides a relationship between the logistic home team winning probability and the home team winning percentage.

It is important to note here that analysts may need to incorporate an adjustment to the spread calculation if the data results are skewed (see Chapter 3 : Probability Models).

The solution to this model is (Table 9.4):

Table 9.4 Logistic Probability Regression: 2015 MLS

Statistic	Value	t-Stat
a_0	−1.409	−2.7
a_1	3.026	4.0
b_0 (Home-field advantage)	0.726	
R^2	4.22%	
F-value	15.623	
Standard Error	1.580	

Therefore, after computing the home team winning probability, the expected spread is estimated from the following equation based on the regression results:

$$Estimated\ Spread = -1.41 + 3.03 \cdot \gamma^*$$

A graph illustrating the estimated spread from the probability estimates is shown in Fig. 9.6.

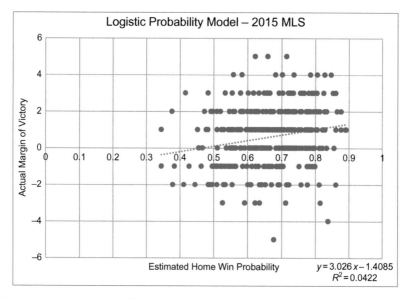

Figure 9.6 Logistic Probability Model: 2015 MLS.

Performance

The logistic probability model's projections were correct in 195 of 357 matches for an adjusted winning percentage of 65.0%. This is an improvement over the Vegas line, which only predicted the winning team 187 times for an adjusted winning percentage of 62.7%.

The away team was the favorite in 25 of the 357 matches, with a maximum win probability of 65.7%; the visiting favorites went 10−12−3 in those 25 games. Home team favorites had win probabilities as high as 89.1%, as did the Red Bulls for a September match against the Chicago Fire, which New York won 3−2. When the win probability was at least 80%, the favorites went 27−4−5, an adjusted winning percentage of 81.9%.

In the playoffs, the logistic probability model's favorites won 12 of the 17 matches while losing 3 with 2 draws.

Rankings

The top eight teams in the logistic ratings—the New York Red Bulls, Portland, Dallas, Los Angeles, Seattle, Columbus, Kansas City, and New England—all made the playoffs. The San Jose Earthquakes at No. 9 were the highest-ranked nonplayoff team, followed by Real Salt Lake at No. 12. The two playoff teams not ranked in the top 12 were DC United (No. 13) and Toronto FC (No. 16). The Portland Timbers, who won the MLS Cup, were ranked No. 2.

Example

Toronto FC ranked 16th among the 20 teams with a logistic rating of 2.3105, while the Red Bulls' rating of 3.1296 led Major League Soccer. Despite the large gap between the two ratings, the logistic probability model favors New York by a relatively small margin, thanks in large part to the home-field advantage parameter:

$$y = \frac{1}{1 + e^{-(2.3105 - 3.1296 + 0.7255)}} = \frac{1}{1 + e^{-0.0936}} = 0.4766$$

We can then take this probability and combine it with the regression parameters b_0 (-1.409) and b_1 (3.027) to estimate that this matchup would result in a very narrow win by Toronto:

$$Estimated\ Spread = -1.409 + 3.027 \cdot 0.4766 \pm SE = 0.034 \pm 1.580$$

$$Probability_{Home} = normcdf(0.034,\ 0,\ 1.580) = 50.86\%$$

9.4 TEAM RATINGS MODEL

The Team ratings prediction model is a linear regression model that uses the team ratings determined from the logistic probability model as the explanatory variables to estimate home team victory margin. This is one of the reasons why the logistic model is one of the more important sports models since its results can be used in different modeling applications.

The team ratings regression has form:

$$Y = b_0 + b_1 \cdot Home\ Rating + b_2 \cdot Away\ Rating + \varepsilon$$

The variables of the model are:

Home Rating = home team strength rating (from logistic probability model)

Away Rating = away team strength rating (from logistic probability model)

b_0 = home-field advantage value

b_1 = home team rating parameter

b_2 = away team rating parameter

Y = home team's victory margin (positive indicates home team is favored and negative value indicates away team is favored)

The probability of winning is determined from:

$$Probability = 1 - NormCDF(0, Y, SE)$$

The betas of this model, b_1 and b_2, are determined from a linear regression analysis as described in Chapter 2, Regression Models.

Regression Results

The best fit regression equation to predict home team victory margin from team strength ratings is:

Estimated Victory Margin $= -0.096 + 0.771 \cdot Home\ Rating - 0.493 \cdot Away\ Rating$

The regression had an R^2 of 4.26% and significant t-Stat for each of the team rating parameters. The standard error of this model is 1.58, indicating that 70% of the time the actual spread will be within 1.58 goals of our predicted spread. The regression results and graph showing the actual victory margin as a function of estimated victory margin are in Table 9.5 and Fig. 9.7, respectively.

Table 9.5 Team Ratings Model: 2015 MLS		
Statistic	**Value**	**t-Stat**
b_0	-0.096	-0.1
b_1 (Home team rating)	0.771	3.3
b_2 (Away team rating)	-0.493	-2.1
R^2	4.26%	
F-value	7.873	
Standard Error	1.582	

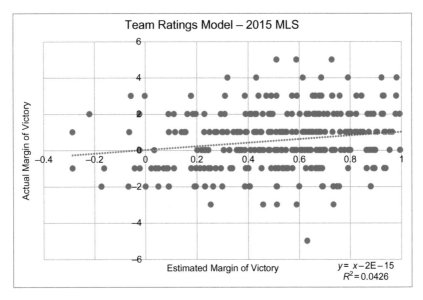

Figure 9.7 Team Ratings Model: 2015 MLS.

Performance

The team ratings model's favorites won 195 of 357 matches for an adjusted success rate of 65.0%, eight wins more than the Vegas line.

Visiting teams were favored in only 13 matches—8 of which were home games for the Chicago Fire—combining for a record of 6−6−1. The average home team's win probability was 64.5%, with a maximum of 82.0% for the Red Bulls when they played Chicago at home. There were 50 matches for which the predicted spread was at least one full goal, equivalent to a win probability of 73.6%, over which the favorites went 35−9−6. In the playoffs, the model's favorites went 12−3−2. In the final, the model gave the Columbus Crew a 60% chance of beating the visiting Portland Timbers, but Portland won the match 3−2.

Example

Toronto FC's rating of 2.3105 was 12th in Major League Soccer in 2015, while the Red Bulls had a league-best 3.1296. The team ratings model equates that to a margin of victory of about one-seventh of a goal:

$$\text{Est. Victory Margin} = -0.096 + 0.771(2.3105) - 0.493(3.1296) = 0.142 \pm 1.582$$

The output of this formula is the estimated victory margin for the home team, and a negative result indicates that the visiting team is favored.

The estimated winning probability is 53.6% for Toronto and 46.4% for the Red Bulls:

$$\text{Probability} = 1 - NormCDF(0, \ 0.142, \ 1.582) = 53.6\%$$

9.5 LOGIT SPREAD MODEL

The logit spread model is a probability model that predicts the home team victory margin based on an inferred team rating metric. The model transforms the home team victory margin to a probability value between 0 and 1 via the cumulative distribution function and then estimates model parameters via logistic regression analysis.

The logit spread model has following form:

$$y = b_0 + b_h - b_a$$

where b_0 denotes a home-field advantage parameter, b_h denotes the home team parameter value, and b_a denotes the away team parameter value.

The left-hand side of the equation y is the log ratio of the cumulative density function of victory margin (see chapter: Sports Prediction Models, for calculation process).

In this formulation, the parameters values b_0, b_h, b_a are then determined via ordinary least squares regression analysis. The results of this analysis are shown in Table 9.1.

Estimating Spreads

Estimating the home team winning margin is accomplished as follows.

If team k is the home team and team j is the away team, we compute y using the logit parameters:

$$y = b_0 + b_k - b_j$$

Compute y^* from y via the following adjustment:

$$y^* = \frac{e^y}{1 + e^y}$$

Compute z as follows:

$$z = norminv(y^*)$$

And finally, the estimated home team spread:

$$Estimated\ Spread = \bar{s} + z \cdot \sigma_s$$

where
   $\bar{s}$ = average home team winning margin (spread) across all games
   σ_s = standard deviation of winning margin across all games

Estimating Probability

The corresponding probability of the home team winning is determined by performing a regression analysis of actual spread as a function of estimated spread to determine a second set of model parameters. This model has form:

$$Actual\ Spread = a_0 + a_1 \cdot Estimated\ Spread$$

To run this regression, we need to compute the estimated spread for all games using the logit spread parameters from above (see Table 9.1).

The solution to this model is (Table 9.6):

Table 9.6 Logit Spread Model: 2015 MLS	
Statistic	**Value**
$\bar{s}$ (Average home victory margin)	0.602
σ_s (Home victory margin standard deviation)	1.612
b_0 (Home-field advantage)	0.024
R^2	5.00%
F-value	18.679
Standard Error	1.574

A graphical illustration of this model is (Fig. 9.8):

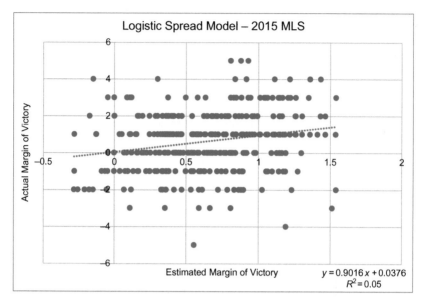

Figure 9.8 Logit Spread Model: 2015 MLS.

Performance

Along with having the best regression statistics, the logit spread model tied for the second-most correct in-sample predictions. Its favorites won 197 out of 357 matches, an adjusted winning percentage of 65.5%. The visiting team was favored in only 22 matches (6.2%). The Chicago Fire was an underdog at home five times, followed by four matches apiece for Orlando City, Philadelphia Union, and Real Salt Lake. The Red Bulls were a road favorite eight times, and also comprised the six highest win probability estimates.

When the home team's win probability was over 70%, they combined for a record of 83−20−15 and an adjusted winning percentage of 76.7%. Favorites of at least 80% went 11−2−0 (84.6%). In the playoffs, the model's favorites won 13 matches against 2 losses and 2 draws.

The logit spread model had the highest R^2 (4.9987%) and lowest standard deviation (1.5738 goals) of the six models when applied to the 2015 Major League Soccer season.

Rankings

The Red Bulls were the top-ranked team by logit spread rating, followed by FC Dallas and the Los Angeles Galaxy. The San Jose Earthquakes were ranked seventh, the highest ranking among nonplayoff teams. At No. 13 the Houston Dynamo was the only other nonplayoff team ranked ahead of a playoff team, as the Dynamo's rating of 2.6343 was just ahead of DC United's 2.3622. The Columbus Crew, winners of the 2015 MLS Cup, ranked ninth.

Example

Toronto FC's logit spread parameter was 2.4005 (12th of the 20 teams), while the Red Bulls led the MLS with a rating of 2.9907. The home-field advantage factor (b_0) was 0.0236.

$$y = \frac{1}{1 + e^{-(2.4005 - 2.9907 + 0.0236)}} = \frac{1}{1 + e^{-0.5666}} = 0.5084$$

According to the logit spread model, Toronto would be a very narrow favorite hosting the Red Bulls, with a win probability between 50% and 51%:

$$Victory\ Margin = z \cdot \sigma_s + \bar{s} = norminv(0.5084) \cdot 1.612 + 0.602 = 0.033$$

$$Probability = 1 - normdist(0,\ 0.033,\ 1.612) = 50.81\%$$

9.6 LOGIT POINTS MODEL

The logit points model is a probability model that predicts the home team victory margin by taking the difference between home team predicted goals and away team predicted goals. The predicted goals are determined based on inferred team "ratings" similar to the logit spread model discussed above.

The logit points model has following form:

$$h = c_0 + c_h - c_a$$

$$a = d_0 + d_h - d_a$$

where

h is the transformed home team goals, c_0 denotes a home-field advantage, c_h denotes the home team rating, and c_a denotes the away team rating corresponding to home team goals.

a is the transformed away team goals, d_0 denotes a home-field advantage, d_h denotes the home team rating, d_a denotes the away team rating corresponding to away team goals.

The left-hand side of the equation h and a is the log ratio of the cumulative density function of home team goals and away team goals respectively (see chapter: Sports Prediction Models, for a description).

Estimating Home and Away Team Goals

Estimating the home team goals is accomplished directly from the home goals team ratings. These rating parameters are shown in Table 9.1. If team k is the home team and team j is the away team, the transformed home team goals are:

$$h = c_0 + c_k - c_j$$

$$h^* = \frac{e^h}{1 + e^h}$$

Then

$$z = norminv(h^*)$$

And finally, the x-value is:

$$Home\ Points = \overline{h} + z \cdot \sigma_h$$

where

$\overline{h}$ = average home team goals
σ_h = standard deviation of home team goals

Away goals are estimated in the same manner but using the team ratings for the away goals model.

Estimating Spread

The estimated home team victory margin is computed directly from the home team goals and away team goals as follows:

$$Est.\ Spread = Home\ Team\ Goals - Away\ Team\ Goals$$

Estimating Probability of Winning

The corresponding probability of winning is determined by performing a regression analysis of actual spread as a function of estimated spread. The model has form:

$$Actual\ Spread = b_0 + b_1 \cdot Est.\ Spread$$

The solution to this model is (Table 9.7):

Table 9.7 Logit Points Model: 2015 MLS		
Statistic	**Home**	**Away**
$\bar{s}$ (Average score)	1.681	1.078
σ_s (Score standard deviation)	1.278	1.016
b_0 (Home-field advantage)	0.078	0.087
R^2	4.94%	
F-value	18.455	
Standard Error	1.574	

A graphical illustration of this model is (Fig. 9.9):

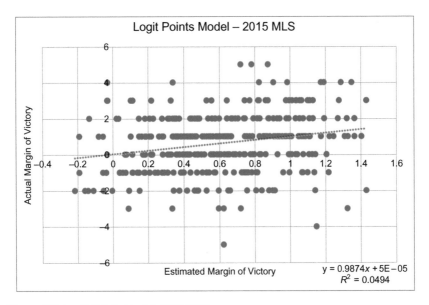

Figure 9.9 Logit Points Model: 2015 MLS.

Performance

The logit points model matched the logit spread model with an adjusted winning percentage of 65.5%, as its favorites won 197 times out of 357 matches. These two models picked different favorites only four times, with each winning twice and losing twice. In terms of projected win probabilities, the two models' outputs were within 3.75 percentage points of each other for every match, and within 3 percentage points for all but five matches.

The logit points model was 12−3−2 in predicting in-sample playoff games. One of the four matches on which the logit points and logit spread models disagreed was a quarterfinal between the Red Bulls and DC United in Washington. Logit points gave DC a 52.8% chance of victory while logit spread called the match ever so slightly for the visiting Red Bulls, calculating a 49.87% probability for DC New York went on to win 1−0.

Example

The model calculated Toronto's ratings to be $b_h = 2.2228$ and $d_h = 2.2146$, with values of $b_a = 2.8652$ and $d_a = 2.2978$ for the Red Bulls and $b_0 = 0.0781$ and $d_0 = 0.0868$ for the home-field advantage factor.

$$Home\ Score = norminv\left(\frac{1}{1 + e^{-(0.0781 + 2.2228 - 2.8652)}}\right) \cdot 1.278 + 1.681 = 1.231$$

$$Away\ Score = norminv\left(\frac{1}{1 + e^{-(0.0868 + 2.2146 - 2.2978)}}\right) \cdot 1.016 + 1.078 = 1.081$$

These scores point to a Toronto victory by less than a quarter of a goal, and a win probability of 53.8%.

$$Victory\ Margin = 1.231 - 1.081 = 0.150$$

$$Probability = 1 - normcdf(0,\ 0.150,\ 1.574) = 53.81\%$$

9.7 EXAMPLE

The six models were unanimous in their conclusions: Toronto FC should defeat the New York Red Bulls. Two models, logistic probability and logit spread, gave the edge to Toronto by less than a 51−49% margin, while the game scores model had Toronto as better than a 2-to-1 favorite and had the only line of over half a goal (Table 9.8).

Table 9.8 Example Results

Model	Favorite	Underdog	Line	P(TOR Win)	P(NYRB Win)
Game Scores	Toronto FC	NY Red Bulls	0.78	68.8%	31.2%
Team Statistics	Toronto FC	NY Red Bulls	0.41	60.2%	39.8%
Logistic Probability	Toronto FC	NY Red Bulls	0.03	50.9%	49.1%
Team Ratings	Toronto FC	NY Red Bulls	0.14	53.6%	46.4%
Logit Spread	Toronto FC	NY Red Bulls	0.03	50.8%	49.2%
Logit Points	Toronto FC	NY Red Bulls	0.15	53.8%	46.2%
Average	Toronto FC	NY Red Bulls	0.26	56.3%	43.7%

9.8 OUT-SAMPLE RESULTS

We performed an out-sample analysis where we predicted our game results using a walk forward approach. In this analysis, we use previous game results data to predict future games. Here, the model parameters were estimated after about 10 games per team and then we predicted the winning team for the next game.

The regression based models using game scores and team statistics were associated with declines of −6.6% and −3.9% respectively. The probability models only had a slight reduction in predictive power with winning percentages consistent with in-sample data (e.g., −1.1%) except for the logit points model, which was down by −4.4%. For all models we found the predictive power of the model to begin to converge to the in-sample results after about 20 games per team (Fig. 9.10).

9.9 CONCLUSION

In this chapter we applied six different sports model approaches to Major League Soccer results for the 2015 season. The models were used to predict winning team, estimated home team victory margin, and probability of winning. As with shootouts during regular-season NHL games, when predicting outcomes some consideration must be made for the large percentage of MLS matches, roughly one in five, that will end in a draw. All of the models had adjusted winning percentages in the neighborhood of 65%, with the team statistics model performing the best at 66.1%, despite the highest regression error and lowest regression R^2 value of the six models. The logit spreads and logit points models had the next-best winning percentage, lowest regression error, and highest regression R^2 values,

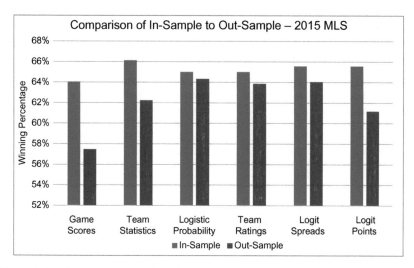

Figure 9.10 Comparison of In-Sample to Out-Sample: 2015 MLS.

followed by the other two probability models, logistic probability and team ratings.

We found that in all cases, these models performed better than the Vegas line using in-sample data and five of six of the models performed better than the Vegas line using out-sample data. The Logit Spread and Logit Points models had the highest goodness of fit (each with $R2 = 65.5\%$) and the lowest regression (each with $seY = 1.57$). These models performed better than the data driven models, but the Team Statistics model had the highest winning percentage at 66.1%. In all cases, our modeling approaches proved to be a valuable predictor of future game results including predicting winning team, winning spread, and probability of winning.

CHAPTER 10

Baseball—MLB

In this chapter, we apply the sports modeling techniques introduced in previous chapters—game scores, team statistics, logistic probability, team ratings, logit spreads, and logit points—to Major League Baseball (MLB) data for the 2015 season.

The models were evaluated in three different ways based on in-sample data: winning percentage, R^2 goodness of the estimated victory margin, and the regression error. Out-sample performance results are discussed at the end of the chapter.

Fig. 10.1 depicts the winning percentage by model. The game scores, team ratings, logistic probability, and logit spread models all picked the winners between 57% and 58.5% of the time, while the team statistics regression's win rate was a bit lower at 55.6%. Fig. 10.2 illustrates the R^2 goodness of fit for each model. All six models had relatively low R^2 values of between 2.4% and 3.6%. Fig. 10.3 shows each model's regression error for predicting the home team's victory margin, all of which were nearly identical, ranging from 4.18 to 4.21.

The regression results for the different models is shown in Table 10.1. These data will be used throughout the chapter. Each of the prediction models is described in the remainder of this chapter.

10.1 GAME SCORES MODEL

The game scores regression model predicts game outcomes based on the average number of runs scored and allowed by each team.

The game scores model has form:

$$Y = b_0 + b_1 \cdot HRS + b_2 \cdot HRA + b_3 \cdot ARS + b_4 \cdot ARA + \varepsilon$$

In this representation, the dependent variable Y denotes the home team's margin of victory (or defeat). A positive value indicates the home team won by the stated number of runs and a negative value indicates the home team lost by the stated number of runs.

Optimal Sports Math, Statistics, and Fantasy.
DOI: http://dx.doi.org/10.1016/B978-0-12-805163-4.00010-4
Copyright © 2017 Robert Kissell and James Poserina.
Published by Elsevier Ltd. All rights reserved.

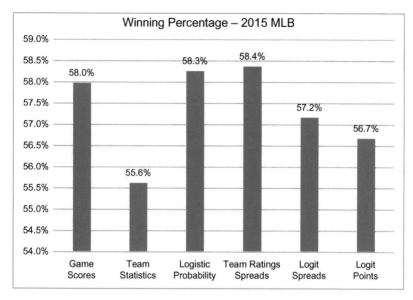

Figure 10.1 Winning Percentage: 2015 MLB.

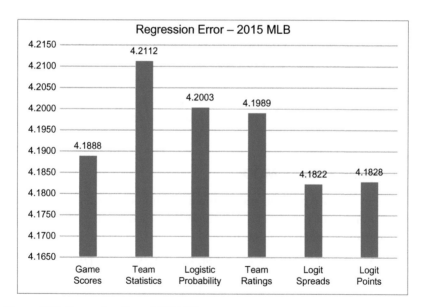

Figure 10.2 Regression Error: 2015 MLB.

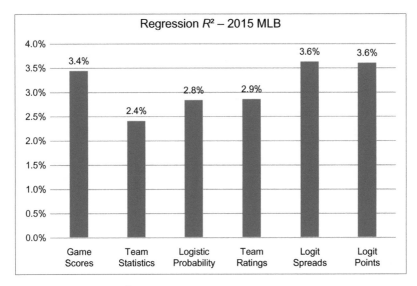

Figure 10.3 Regression R^2: 2015 MLB.

The variables of the model are:

Y = home team victory margin
HRS = home team average runs scored per game
HRA = home team average runs allowed per game
ARS = away team average runs scored per game
ARA = away team average runs allowed per game
b_0 = home-field advantage value
b_1 = home team runs scored parameter
b_2 = home team runs allowed parameter
b_3 = away team runs scored parameter
b_4 = away team runs allowed parameter

The betas of this model are determined from a linear regression analysis as described in Chapter 2, Regression Models.

This model can also be used to calculate the probability p that the home team will win the game as follows:

$$p = 1 - NormCDF(0, Y, SeY)$$

Here, Y and SeY are the expected victory margin and regression error term respectively, and zero indicates the reference point used for the calculation. That is, the probability that the winning margin will be greater than zero.

Table 10.1 Regression Results and Rankings

| | | | Team Statistics | | | | Ratings | | | | Rankings | | | | |
Team	W – L	Pct	RF/G	RA/G	OBP	Opponent OBP	Logistic Rating	Logit Spread	Logit Home Points	Logit Away Points	Game Scores	Team Statistics	Logistic Ratings	Logit Spread	Logit Points
Arizona Diamondbacks	79 – 83	.488	4.44	4.40	.324	.322	2.338	2.384	2.520	2.696	15	14	22	22	22
Atlanta Braves	67 – 95	.414	3.54	4.69	.314	.339	2.020	1.888	1.980	2.901	30	27	29	29	30
Baltimore Orioles	81 – 81	.500	4.40	4.28	.307	.321	2.587	2.736	2.981	2.584	14	22	14	5	5
Boston Red Sox	78 – 84	.481	4.62	4.65	.325	.327	2.511	2.664	2.874	2.640	16	16	17	9	8
Chicago Cubs	97 – 65	.599	4.25	3.75	.321	.290	2.819	2.579	2.525	2.430	6	2	5	14	15
Chicago White Sox	76 – 86	.469	3.84	4.33	.306	.322	2.520	2.404	2.283	2.443	25	24	16	21	21
Cincinnati Reds	64 – 98	.395	3.95	4.65	.312	.328	2.076	2.136	2.350	2.860	28	23	28	27	27
Cleveland Indians	81 – 80	.503	4.16	3.98	.325	.297	2.630	2.680	2.911	2.633	12	4	11	6	6
Colorado Rockies	68 – 94	.420	4.55	5.21	.315	.353	2.084	2.116	2.699	3.293	26	29	27	28	28
Detroit Tigers	74 – 87	.460	4.28	4.99	.328	.330	2.482	2.319	2.209	2.463	27	15	18	23	23
Houston Astros	86 – 76	.531	4.50	3.81	.315	.299	2.689	2.910	2.670	2.055	3	7	8	2	2
Kansas City Royals	95 – 67	.586	4.47	3.96	.322	.314	2.990	2.852	2.716	2.194	5	12	1	3	4
Los Angeles Angels	85 – 77	.525	4.08	4.17	.307	.313	2.662	2.598	2.381	2.235	19	19	10	12	11
Los Angeles Dodgers	92 – 70	.568	4.12	3.67	.326	.297	2.620	2.510	2.359	2.372	11	3	12	19	19
Miami Marlins	71 – 91	.438	3.78	4.19	.310	.322	2.110	2.188	2.235	2.671	22	21	26	25	24
Milwaukee Brewers	68 – 94	.420	4.04	4.55	.307	.329	2.168	2.212	2.385	2.844	24	26	25	24	25
Minnesota Twins	83 – 79	.512	4.30	4.32	.305	.321	2.666	2.597	2.532	2.405	17	25	9	13	13
New York Mets	90 – 72	.556	4.22	3.78	.312	.296	2.592	2.544	2.426	2.322	10	8	13	16	14
New York Yankees	87 – 75	.537	4.72	4.31	.323	.316	2.710	2.851	2.803	2.214	7	13	7	4	3
Oakland Athletics	68 – 94	.420	4.28	4.50	.312	.316	2.275	2.546	2.412	2.366	20	18	23	15	17
Philadelphia Phillies	63 – 99	.389	3.86	4.99	.303	.341	1.931	1.884	2.052	2.946	29	30	30	30	29
Pittsburgh Pirates	98 – 64	.605	4.30	3.68	.323	.311	2.839	2.609	2.475	2.333	4	9	4	11	12
San Diego Padres	74 – 88	.457	4.01	4.51	.300	.321	2.223	2.171	2.176	2.654	23	28	24	26	26
San Francisco Giants	84 – 78	.519	4.30	3.87	.326	.305	2.443	2.526	2.348	2.301	8	5	20	18	16
Seattle Mariners	76 – 86	.469	4.05	4.48	.311	.322	2.448	2.417	2.200	2.360	21	20	19	20	20
St. Louis Cardinals	100 – 62	.617	3.99	3.24	.321	.310	2.887	2.668	2.534	2.308	2	10	2	8	9
Tampa Bay Rays	80 – 82	.494	3.98	3.96	.314	.304	2.558	2.678	2.629	2.381	18	11	15	7	7
Texas Rangers	88 – 74	.543	4.64	4.52	.325	.328	2.717	2.637	2.746	2.541	13	17	6	10	10
Toronto Blue Jays	93 – 69	.574	5.50	4.14	.340	.304	2.843	3.165	2.933	1.943	1	1	3	1	1
Washington Nationals	83 – 79	.512	4.34	3.92	.321	.301	2.389	2.528	2.489	2.445	9	6	21	17	18
Home-Field Advantage							0.172	0.002	0.167	0.167					0.167

Regression Results

The best fit regression equation for predicting the victory margin as a function of team runs scored and runs allowed is below. The resulting victory margin is positive when the home team is favored and negative when the visiting team is favored.

This model equation is:

$$Est. \; Victory \; Margin = -2.163 + 1.161 \cdot HRS - 0.772 \cdot HRA$$
$$- 0.846 \cdot ARS + 1.009 \cdot ARA$$

The game scores model yielded an R^2 value of 3.4%. The t-Stats were -11.68 and 7.74 for visiting teams' runs scored and runs allowed, and 13.72 and -8.89 for home teams, and an F-value of 101.55.

The signs of the input variables are intuitive. The sign of the sensitivity parameter is expected to be positive for home runs scored (HRS) and away runs allowed (ARA). In both of these cases, the home team will score more runs if these input values are higher. Similarly, the signs of the sensitivity parameters for home runs allowed (HRA) and away team runs scored (ARS) are expected to be negative. The home team is expected to win by fewer runs or possibly lose if the HRA and/or ARS increase in value.

A graph showing the actual spreads as a function of estimated spreads is shown in Fig. 10.4 (Table 10.2).

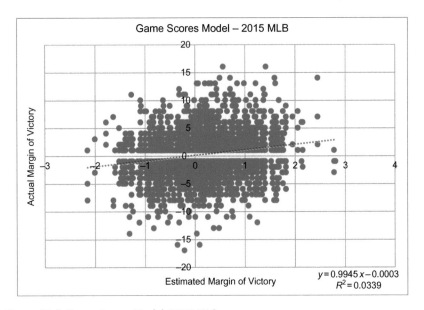

Figure 10.4 Game Scores Model: 2015 MLB.

Table 10.2 Game Scores Model: 2015 MLB		
Statistic	Value	t-Stat
b_0	− 2.163	− 1.2
b_1 (Home runs scored)	1.161	4.9
b_2 (Home runs allowed)	− 0.772	− 4.0
b_3 (Visitor runs scored)	− 0.845	− 3.6
b_4 (Visitor runs allowed)	1.009	5.2
R^2	3.44%	
F-value	21.883	
Standard Error	4.189	

Performance

The game scores model correctly identified the winner in 1429 of 2465 games throughout the regular season and the playoffs, a rate of 58.0%.

The average estimated win probability for the favorite was very low at just 56.1%. In only three matchups was the probability estimated to be over 70%, all of which were home games for the Blue Jays: a two-game series with the Philadelphia Phillies in July (74.9%, series was split), three games with the Atlanta Braves in April (74.6%, Braves won two of three), and a three-game set with the Detroit Tigers in August (72.1%, Blue Jays swept). In 288 games the win probability was between 50% and 51%, and overall the median was 55.4%. When the probability was 60% or greater, the model was accurate 63.1% of the time.

In the postseason, only three home teams were underdogs, the Texas Rangers in the American League Division Series (ALDS) and the Kansas City Royals in the American League Championship Series (ALCS), both against the Blue Jays, and the Cubs at home in the National League Division Series against the St. Louis Cardinals. The Rangers lost those two games, while the Royals won their three games and the Cubs won both of theirs. At 65.6% the biggest favorites were the Blue Jays at home against Texas in the ALDS; Toronto would lose two of the three games. The model was over 50% accurate in both American League Division Series and the World Series, but not in the either of the National League Division Series, either League Championship Series, or either wild card game.

Rankings

The top spot of the game scores rankings went to the Toronto Blue Jays, who won the American League East, by virtue of their 5.50 runs scored per game, a full run per game more than all but four other teams. Their average run differential of +1.364 led the majors by an extremely wide margin. The St. Louis Cardinals were second in average run differential (+0.753) and in the rankings, as they were the only team to win 100 games during the 2015 regular season. The Kansas City Royals, who won the World Series, were ranked fifth, while their counterparts in the Fall Classic, the New York Mets, placed tenth. The 10 playoff teams finished in the top 13 along with 3 nonplayoff teams, the No. 8 San Francisco Giants, No. 9 Washington Nationals, and No. 12 Cleveland Indians. The Texas Rangers were the lowest-ranked playoff team at No. 13. Along with top 10 teams the Mets and Nationals, the National League East also comprised the bottom two, with the Philadelphia Phillies (63−99) at No. 29 and the Atlanta Braves (67−95) at No. 30.

Example
We will use the six models to predict the outcome of a game between the Chicago Cubs (97−65) and the Toronto Blue Jays (93−69), the runners-up in 2015's two League Championship Series, if they were to play north of the border. The Cubs scored 4.25 runs per game while allowing 3.75, while the Blue Jays scored 5.50 runs a game and gave up 4.14. The game scores model calculates a Blue Jays victory by a little over one run with a win probability of 61.5%:

$$Est.\ Victory\ Margin = 2.163 + 1.161(5.50) - 0.772(4.14) - 0.846(4.25)$$
$$+ 1.009(4.14) = 1.226 \pm 4.189$$

$$p = 1 - NormCDF(0, 1.226, 4.189) = 61.511\%$$

10.2 TEAM STATISTICS MODEL

The team statistics regression uses measures of team performance to predict game results.

In this section, we will demonstrate the regression using on-base percentage (OBP), both the teams' own as well as that of their opponents. On-base percentage is the sum of hits, walks (intentional or otherwise), and hit batsmen divided by the sum of at-bats, walks, hit batsmen, and sacrifice flies. Sacrifice flies do not count against batting average, but they do count against OBP.

$$OBP = \frac{H + BB + HBP}{AB + BB + HBP + SF}$$

Table 10.3 Team Statistics Model: 2015 MLB		
Statistic	**Value**	**t-Stat**
b_0	−3.654	−0.6
b_1 (Home on-base percentage)	24.306	2.4
b_2 (Home opponent on-base percentage)	−18.757	−3.0
b_3 (Away on-base percentage)	−20.570	−2.0
b_4 (Away opponent on-base percentage)	27.167	4.3
R^2		2.40%
F-value		15.147
Standard Error		4.211

The team statistics linear regression model has form:

$$Y = b_0 + b_1 \cdot HOBP + b_2 \cdot HOBP_{Opp} + b_3 \cdot AOBP + b_4 \cdot AOBP_{Opp} + \varepsilon$$

The variables of the model are:

Y = outcome value we are looking to predict
$HOBP$ = home team's on-base percentage
$HOBP_{Opp}$ = home team's opponents' on-base percentage
$AOBP$ = away team's on-base percentage
$AOBP_{Opp}$ = away team's opponents' on-base percentage
b_0, b_1, b_2, b_3, b_4 = model parameter, sensitivity to the variable
ε = model error

The betas of this model are determined from a linear regression analysis as described in Chapter 2, Regression Models. The results of these model are shown in Table 10.3.

Regression Results

The result of the regression model for predicting home team victory margin from our team statistics is:

$$Est.\ Victory\ Margin = -3.654 + 24.306(HOBP) - 18.757\left(HOBP_{Opp}\right)$$
$$- 20.570(AOBP) + 27.167(AOBP_{Opp}) \pm 4.211$$

All four parameters had significant t-statistics, the highest being 4.3 for away team's opponents' OBP (Fig. 10.5).

Performance

The team statistics model had an accuracy of 55.6%, identifying the winners in 1371 of 2465 games in the regular season and in the postseason.

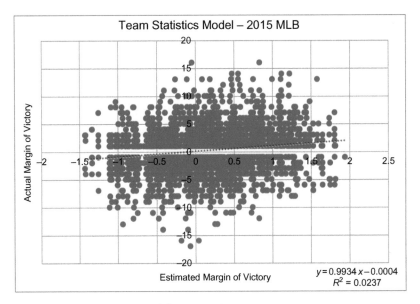

Figure 10.5 Team Statistics Model: 2015 MLB.

The model was 51.9% correct when the visiting team was favored, and 58.0% when the home team was given a higher win probability. The highest win probability was 67.75% for the Blue Jays when they hosted the Phillies in July, and conversely the highest probability for a visiting team was also for the Blue Jays (63.28%) when they traveled to Philadelphia. There were 86 games for which the home team was favored by at least 62.75%, and in all of them the visiting team was either the Phillies, the Colorado Rockies, or the Atlanta Braves. When the win probability was at least 60%, the model's accuracy was 65.5%, compared with an average win probability of 62.0% across that sample.

In the postseason the model's predictions were 41.7% accurate. It was over 50% for only two series, the National League Division Series between St. Louis and Chicago, and the American League Division Series between Texas and Toronto. The Cubs were favored in each game of the NLCS, in which they were swept by the Mets. The Mets, in turn, were favored in each game of the World Series, which the Kansas City Royals won in five games. The margins in the World Series were extremely narrow: 51/49 for Games 1 and 2 in Kansas City, and 53/47 for Games 3—5 in New York. When the model's win probabilities for all of the winners are averaged together, it works out to a mean of 49.6%.

Rankings

The Blue Jays took the top ranking with a rating of 1.885 thanks largely to their MLB-best .340 OBP, 12 points higher than any other club. On the other side of the ball, the Cubs held their opponents to a .290 OBP, which helped them rise to No. 2 with a rating of 1.478. Each team in the top eight held the opposition to an OBP of no more than .305. The lowest-ranked playoff team was the Texas Rangers at No. 17, while the Cleveland Indians (.325 gained/.297 allowed), third in the AL Central with a record of just 81−80, ranked No. 4, best among nonplayoff teams. The National League champion Mets ranked at No. 8, between the Astros and Pirates, while the World Series champion Royals placed 12th.

Example

The Toronto Blue Jays led the major leagues with a .340 team OBP, while holding their opponents to .304, seventh best overall. The Chicago Cubs' OBP of .321 was toward the middle of the pack, but their pitching and defense kept the opposition to an OBP of .290, the best in baseball. So our matchup has the team with the highest OBP and the team that allowed the lowest.

$$\text{Est. Margin of Victory} = -3.654 + 24.306(.33963) - 18.757(.30368)$$
$$- 20.570(.32139) + 27.167(.29015) \pm SE$$
$$= 0.1764 \pm 4.211$$

This spread works out to a 51.7% probability of a Toronto victory:

$$p = 1 - NormCDF(0, 0.1764, 4.211) = 51.670\%$$

10.3 LOGISTIC PROBABILITY MODEL

The logistic probability model infers a team strength rating based only on game outcomes such as whether the team won, lost, or tied the game. The result of the game is determined from the perspective of the home team, but analysts can use the same approach from the perspective of the visiting team.

The logistic model is as follows:

$$y^* = \frac{1}{1 + \exp\left\{-(b_0 + b_h - b_a)\right\}}$$

Here, b_0 denotes a home-field advantage parameter, b_h denotes the home team rating parameter value, and b_a denotes the away team rating parameter value. The value y^* denotes the probability that the home team will win the game. Team ratings for the logistic probability are determined via maximum likelihood estimates and are shown in Table 10.1.

Estimating Spread

The estimated spread (i.e., home team victory margin) is determined via a second analysis where we regress the actual home team spread on the estimated probability y^* (as the input variable). This regression has form:

$$Actual\ Spread = a_0 + a_1 \cdot y^*$$

This model now provides a relationship between the logistic home team winning probability and the home team winning percentage. It is important to note here that analysts may need to incorporate an adjustment to the spread calculation if the data results are skewed (see Chapter 3: Probability Models).

The solution to this model is (Table 10.4):

Table 10.4 Logistic Probability Regression: 2015 MLB

Statistic	Value	t-Stat
a_0	− 4.231	− 8.0
a_1	8.164	8.5
b_0 (Home-field advantage)	0.172	
R^2	2.85%	
F-value	72.279	
Standard Error	4.199	

Therefore, after computing the home team winning probability y^*, the expected winning spread is estimated from the following equation using the regression results:

$$Estimated\ Spread = -4.231 + 8.164 \cdot y^*$$

A graph illustrating the estimated spread from the probability estimates is shown in Fig. 10.6.

Performance

The logistic probability model identified the winner in 1436 of 2465 games, a rate of 58.3%. The biggest favorite was the Cardinals (67.77%) at home

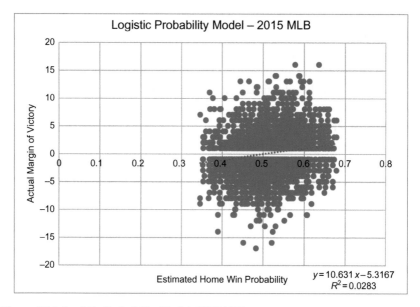

Figure 10.6 Logistic Probability Model: 2015 MLB.

against the Phillies in April, a series in which St. Louis won three out of four by a combined score of 26−14. Naturally, the biggest underdog among home teams was Philadelphia when they hosted the Cardinals in June; St. Louis won the first two games 12−4 and 10−1 before Philadelphia salvaged the third game with a 9−2 victory. When a team was calculated as having a win probability of 61% or more, it won 70.3% of the time. The most even matchup according to this model was a two-game series between the Rangers and the Dodgers in Los Angeles, in which the Dodgers were favored by a margin of 50.02% to 49.98% and in which each team won one game.

Rankings

The World Series champion Royals had the highest rating of 2.9897, followed by the Cardinals (2.8875), Blue Jays (2.8434), Pirates (2.8434), and Cubs (2.8195). The Royals' opponents in the Fall Classic ranked 13th as the Mets drew a rating of 2.5921, lowest among playoff teams. The Twins, Angels, and Indians were the highest ranked among those who stayed home in October, ranking 7th, 8th, and 9th respectively. The bottom seven teams were all National League clubs, with three of the bottom five (No. 26 Marlins, No. 29 Braves, and No. 30 Phillies) representing the NL East. The Oakland Athletics, ranked 23rd, were the only American League team among the bottom 11.

Example

Our matchup features the teams in the No. 3 and No. 5 slots in the logistic probability model's rankings, the Blue Jays (2.8434) and the Cubs (2.8195).

$$y^* = \frac{1}{1 + e^{-(0.1721+2.8434-2.8195)}} = \frac{1}{1 + e^{-0.1961}} = 0.5489$$

We can then take this value and combine it with the regression parameters b_0 (−4.231) and b_1 (8.164) to estimate a margin of 0.250 runs in favor of Toronto. That works out to a win probability of 52.37% for the Blue Jays and 47.63% for the Cubs.

$$Estimated\ Spread = -4.231 + 8.164 \cdot 0.5489 \pm SE = 0.2497 \pm 4.199$$

$$Probability = 1 - normdist(0, 0.2497, 4.199) = 52.37\%$$

10.4 TEAM RATINGS MODEL

The team ratings prediction model is a linear regression model that uses the team ratings determined from the logistic probability model as the explanatory variables to estimate home team victory margin. This is one of the reasons why the logistic model is one of the more important sports models since its results can be used in different modeling applications.

The team ratings regression has form:

$$Y = b_0 + b_1 \cdot Home\ Rating + b_2 \cdot Away\ Rating + \epsilon$$

The variables of the model are:

> $Home\ Rating$ = home team strength rating (from logistic probability model)
>
> $Away\ Rating$ = away team strength rating (from logistic probability model)
>
> b_0 = home-field advantage value
> b_1 = home team rating parameter
> b_2 = away team rating parameter
> Y = value we are looking to predict

The probability of winning is determined from:

$$Probability = 1 - NormCDF(0, Y, SE)$$

The betas of this model, b_1 and b_2, are determined from a linear regression analysis as described in Chapter 2, Regression Models.

Regression Results

The best fit regression equation to predict home team victory margin from team strength ratings is:

$$Victory\ Margin = 0.340 + 1.930 \cdot Home\ Rating - 1.990 \cdot Away\ Rating$$

The regression results and graph showing the actual victory margin as a function of estimated victory margin are in Table 10.5 and Fig. 10.7, respectively.

Table 10.5 Team Ratings Model: 2015 MLB		
Statistic	**Value**	**t-Stat**
b_0	0.340	0.3
b_1 (Home team rating)	1.930	6.3
b_2 (Away team rating)	− 1.990	− 6.5
R^2	2.83%	
F-value	35.847	
Standard Error	4.200	

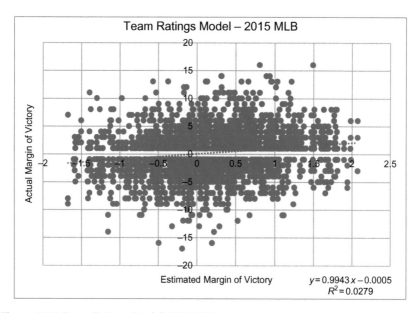

Figure 10.7 Team Ratings Model: 2015 MLB.

Performance

Teams favored by the team ratings model won 1440 of 2465 games, a rate of 58.4%, the highest rate among the six models. When the favorite had a win probability of 61% or more, that team won 70.4% of the time. The favorites' win probabilities averaged 55.5%; between 50% and the average the win rate was 55.1%, and between the average and maximum it was 63.1%.

The most lopsided contests involved the Cardinals (68.89%), Blue Jays (68.17%), Pirates (68.10%), and Cubs (67.78%) when each hosted the Phillies, along with the Royals (68.12%). The favorites won 9 of the 14 games among that sample, even after the Phillies swept the Cubs three straight. The most even matchup was a three-game series in May between the Indians (50.07%) and the Rangers (49.93%) in Cleveland, with Texas winning two out of three.

Example

Toronto's rating of 2.843 was third in the major leagues, behind only the Royals and Cardinals. The Cubs were fifth with a rating of 2.819. With a home-field advantage factor of 0.340, this model projects a win probability of 52.06% for Toronto and 47.94% for the Cubs, with a margin of victory of 0.217 runs in favor of the hometown Blue Jays:

$$Est.\ Victory\ Margin = 0.340 + 1.930(2.843) - 1.990(2.819) \pm SE$$
$$= 0.217 \pm 4.200$$

$$Probability = 1 - NormCDF(0, 0.217, 4.200) = 52.06\%$$

10.5 LOGIT SPREAD MODEL

The logit spread model is a probability model that predicts the home team victory margin based on an inferred team rating metric. The model transforms the home team victory margin to a probability value between 0 and 1 via the cumulative distribution function and then estimates model parameters via logit regression analysis.

The logit spread model has following form:

$$y = b_0 + b_h - b_a$$

where b_0 denotes a home-field advantage parameter, b_h denotes the home team parameter value, and b_a denotes the away team parameter value.

The left-hand side of the equation y is the log ratio of the cumulative density function of victory margin (see chapter: Sports Prediction Models, for calculation process).

In this formulation, the parameters values b_0, b_h, b_a are then determined via ordinary least squares regression analysis. The results of this analysis are shown in Table 10.1.

Estimating Spreads

Estimating the home team winning margin is accomplished as follows.

If team k is the home team and team j is the away team, we compute y using the logit parameters:

$$y = b_0 + b_k - b_j$$

Compute y^* from y via the following adjustment:

$$y^* = \frac{e^y}{1 + e^y}$$

Compute z as follows:

$$z = norminv(y^*)$$

And finally, the estimated home team spread:

$$Estimated\ Spread = \bar{s} + z \cdot \sigma_s$$

where

$\bar{s}$ = average home team winning margin (spread) across all games

σ_s = standard deviation of winning margin across all games

Estimating Probability

The corresponding probability of the home team winning is determined by performing a regression analysis of actual spread as a function of estimated spread to determine a second set of model parameters. This model has form:

$$Actual\ Spread = a_0 + a_1 \cdot Estimated\ Spread$$

To run this regression, we need to compute the estimated spread for all games using the logit spread parameters from above (see Table 10.1).

The solution to this model is (Table 10.6):

Table 10.6 Logit Spread Model: 2015 MLB	
Statistic	Value
$\bar{s}$ (Average home victory margin)	0.190
σ_s (Home victory margin standard deviation)	4.259
b_0 (Home-field advantage)	0.002
R^2	3.62%
F-value	92.590
Standard Error	4.182

A graphical illustration of this model is (Fig. 10.8):

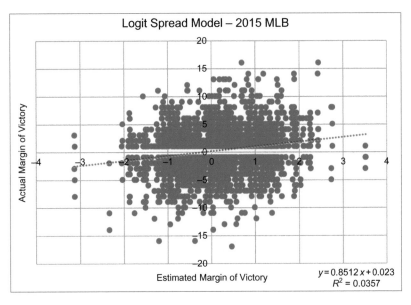

Figure 10.8 Logit Spread Model: 2015 MLB.

Performance

The logit spread model's predictions had an accuracy of 57.2%, as its favorite won 1409 of 2465 games. This rate was very close to 57.3%, the average win probability for the favorite. This model also had the highest projected win probabilies, with a maximum of 79.998% for the Blue Jays when hosting the Phillies. There were 46 games with a win probability over 70%, of which the favorites won 76.1%. The most even matchup was when the Miami Marlins traveled to Colorado in June for a three-game series in which the Rockies were favored by a margin of 50.014% to 49.986% and lost two of three.

In the postseason the model's accuracy was 47.2%. It did pick the Royals both at home (59.6%) and on the road (55.9%) in the World Series. It also favored the Cardinals in each game of the NL Division Series against the Cubs, although for Games 3 and 4 in Chicago, both of which went to the Cubs, St. Louis was given only a 50.4% chance of winning. The average win probability for a favorite in the postseason was 55.2%.

Rankings

The Blue Jays led the rankings with a rating of 3.1645, followed by the Astros (2.9100), Royals (2.8517), and Yankees (2.8512). The next three spots went to nonplayoff teams from the American League, the Orioles, Indians, and Rays, followed by the NL Central champion Cardinals at No. 8. The Nationals at No. 17 were the highest-ranked nonplayoff team from the Senior Circuit. The four lowest-ranked playoff teams were all from the National League, i.e., the Pirates (No. 11), Cubs (No. 14), Mets (No. 16), and Dodgers (No. 19), as were the seven lowest-ranked teams overall, as the Braves (1.8884) and Phillies (1.8844) brought up the rear as the only teams with ratings less than 2.

Example

The average score differential for the Blue Jays at home was $+1.77$, while the friendly confines of Wrigley Field were somewhat less hospitable for the Cubs, as they averaged a $+0.188$ run differential in their home games. The margins for the two teams were much more even when playing on the road, where the Blue Jays averaged $+0.721$, just more than the Cubs' 0.686. This should equate to a much higher logit spread rating for Toronto, and indeed that is the case. The Blue Jays led the major leagues with a value of 3.1645, while the Cubs were 14th overall with a rating of 2.5791. The home-field advantage factor, b_0, was very low at 0.001911.

$$y = \frac{1}{1 + e^{-(3.1645 - 2.5791 - 0.001911)}} = \frac{1}{1 + e^{-0.5835}} = 0.6428$$

In 2015, the average MLB game ended with a home-team victory by 0.1903 runs, with a standard deviation of 4.259. With those parameters, the logit spread model calculates that the Blue Jays should defeat the Cubs at home by 1.75 runs with a win probability of 66.21%:

$$\textit{Victory Margin} = z \cdot \sigma_s + \bar{s} = norminv(0.6428) \cdot 4.259 + 0.1903 = 1.7484$$

$$\textit{Probability} = normdist(1.7484, 4.259) = 66.21\%$$

10.6 LOGIT POINTS MODEL

The logit points model is a probability model that predicts the home team victory margin by taking the difference between home team predicted runs and away team predicted runs. The predicted runs are determined based on inferred team "ratings" similar to the logit spread model discussed previously.

The logit points model has following form:

$$h = c_0 + c_h - c_a$$
$$a = d_0 + d_h - d_a$$

where

h is the transformed home team runs, c_0 denotes a home-field advantage, c_h denotes the home team rating, and c_a denotes the away team rating corresponding to home team runs.

a is the transformed away team runs, d_0 denotes a home-field advantage, d_h denotes the home team rating, d_a denotes the away team rating corresponding to away team runs.

The left-hand side of the equation h and a is the log ratio of the cumulative density function of home team runs and away team runs respectively (see chapter: Sports Prediction Models, for a description).

Estimating Home and Away Team Runs

Estimating the home team runs is accomplished directly from the home runs team ratings. These rating parameters are shown in Table 10.1. If team k is the home team and team j is the away team, the transformed home team runs are:

$$h = c_0 + c_k - c_j$$
$$h^* = \frac{e^h}{1 + e^h}$$

Then

$$z = norminv(h^*)$$

And finally, the x-value is:

$$Home\ Points = \bar{h} + z \cdot \sigma_h$$

where

$$\bar{h} = average\ home\ team\ runs$$
$$\sigma_h = standard\ deviation\ of\ home\ team\ runs$$

Away runs are estimated in the same manner but using the team ratings for the away runs model.

Estimating Spread

The estimated home team victory margin is computed directly from the home team runs and away team runs as follows:

$$Est.\ Spread = Home\ Team\ Points - Away\ Team\ Points$$

Estimating Probability of Winning

The corresponding probability of winning is determined by performing a regression analysis of actual spread as a function of estimated spread. The model has form:

$$Actual\ Spread = b_0 + b_1 \cdot Est.\ Spread$$

The solution to this model is (Table 10.7):

Table 10.7 Logit Points Model: 2015 MLB		
Statistic	**Home**	**Away**
$\bar{s}$ (Average score)	0.180	0.517
σ_s (Score standard deviation)	1.018	0.094
b_0 (Home-field advantage)	0.167	0.167
R^2	3.60%	
F-value	91.905	
Standard Error	4.183	

A graphical illustration of this model is shown in Fig. 10.9.

Performance

The logit points model identified the winner in 1397 of 2465 games. Its success rate of 56.7% was fifth best among the six models, and was below the average win probability of 57.8%.

This model calculated the only win probabilities above 80% for games in two series played in Toronto involving the Phillies (Blue Jays favored 81.4% to 18.6%) and the Braves (Blue Jays 81.7%). Philadelphia split the two games, while Atlanta took two of three. The closest matchup

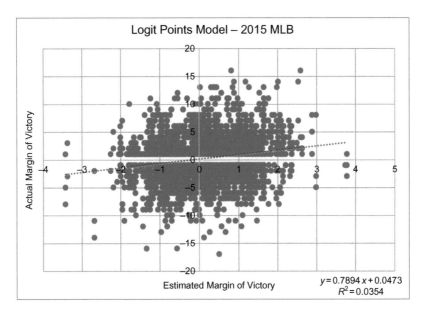

Figure 10.9 Logit Points Model: 2015 MLB.

involved the Chicago White Sox when they played in Detroit, as the Tigers were favored by a margin of 50.005% to 49.995%. They met 10 times over three series in April, June, and September, Detroit winning six of the 10 games.

In the postseason, the model projected the winner 47.2% of the time. This model narrowly favored the home team in each game of the National League Championship Series, won by the Mets in four straight games. New York's win probability was 52.0% in Games 1 and 2, and Chicago's was 51.5% for Games 3 and 4. The Blue Jays were a much stronger favorite in the ALCS, with odds of 60.2% at home and 56.8% on the road in Kansas City; the Royals won 4 games to 2. The model preferred Kansas City in each game of the 2015 World Series, with win probabilities of 59.2% when the Royals were at home and 55.6% for Games 3–5 played in New York.

Example

The logit points model requires that we first estimate the score for our matchup before determining the win probability. Toronto's ratings were $b_h = 2.9334$ and $d_h = 1.9428$, as the Cubs' figures were $b_a = 2.5249$ and

$d_a = 2.4304$. The home-field advantage factors were $b_0 = 0.1672$ and $d_0 = 0.1674$.

$$Home\ Score = norminv\left(\frac{1}{1 + e^{-(0.1672 + 2.9334 - 2.4304)}}\right) \cdot 3.0084 + 4.3469 = 5.4261$$

$$Away\ Score = norminv\left(\frac{1}{1 + e^{-(0.1674 + 1.9428 - 2.4304)}}\right) \cdot 3.1116 + 4.1566 = 3.5332$$

When these estimates are rounded it yields a prediction of Blue Jays 5, Cubs 4, although without rounding the victory margin is closer to two runs than one:

$$Victory\ Margin = 5.4261 - 3.5332 = 1.8929$$

The normal cumulative distribution function yields the probability of a win for the Jays, given the regression's standard deviation of 4.1828. According to the logit points model, that probability is 67.5%, the highest among the six models:

$$Probability = 1 - normcdf(0, 1.8929, 4.1828) = 67.46\%$$

10.7 EXAMPLE

All six models favor the Toronto Blue Jays at home over the Chicago Cubs. Three models gave the Blue Jays around a 52% chance of winning with a margin of around a quarter of a run, while two others put Toronto's chances at 67% giving them nearly two full runs (Table 10.8).

The Cubs and Blue Jays did not face each other during the regular season or in the playoffs in either 2015 or 2016. They last played in Toronto in September 2014, when the Blue Jays swept a very different Cubs team in a three-game series by a combined score of $28-3$.

Table 10.8 Example Results

Model	Favorite	Underdog	Line	P(TOR Win)	P(CHC Win)
Game Scores	Toronto	Chicago Cubs	1.23	61.5%	38.5%
Team Statistics	Toronto	Chicago Cubs	0.18	51.7%	48.3%
Logistic Probability	Toronto	Chicago Cubs	0.25	52.4%	47.6%
Team Ratings	Toronto	Chicago Cubs	0.22	52.1%	47.9%
Logit Spread	Toronto	Chicago Cubs	1.75	66.2%	33.8%
Logit Points	Toronto	Chicago Cubs	1.89	67.5%	32.5%
Average	Toronto	Chicago Cubs	0.92	58.5%	41.5%

10.8 OUT-SAMPLE RESULTS

We performed an out-sample analysis where we predicted our game results using a walk forward approach. In this analysis, we use previous game results data to predict future games. The model parameters were estimated after about 20 games per team. We then predicted the winning team for the next game using this data.

The regression-based models using game scores and team statistics were associated with declines of −6.6% and −3.9% respectively. The probability models only had a slight reduction in predictive power with winning percentages consistent with in-sample data (e.g., −1.1%) except for the logit points model, which was down by −4.4%. For all models we found the predictive power of the model to begin to converge to the in-sample results after about 40 games per team.

Compared to the other sports, MLB required the largest number of games before our models become statistically predictive (approximately 20 games per team) and MLB also required the largest number of games before the out-sample began to converge to in-sample results (approximately 40 games per team). The large number of games required for MLB is most likely due to baseball teams playing series where they play the same team consecutive times in a row thus needing more games to have a large enough sample across different pairs of teams. Additionally, baseball results are highly dependent upon the pitcher, who can only compete about every fifth game, thus again requiring a larger number of games than other sports to be statistically predictive (Fig. 10.10).

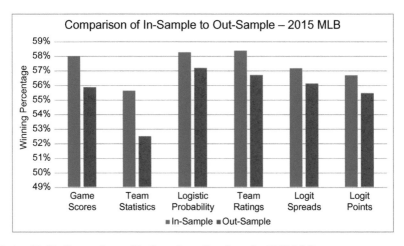

Figure 10.10 Comparison of In-Sample to Out-Sample: 2015 MLB.

10.9 CONCLUSION

In this chapter we applied six different sports model approaches to Major League Baseball results for the 2015 season. The models were used to predict winning team, estimated home team victory margin, and probability of winning. Baseball is somewhat trickier to predict than other sports. The gap between the best and worst teams is much narrower in baseball (a baseball team going 111−51 or 41−121 would almost be historic while an NFL team going 11−5 or 4−12 is rather ordinary), and with games played nearly every day, a star player is more likely to have a day off than in other sports that do not play every day. And perhaps the most significant factor is one unique to baseball, mentioned in Earl Weaver's old aphorism that "momentum is the next day's starting pitcher." Underdogs have a better chance of winning a baseball game than in any other sport, and our analyses bear this out. The best-performing models—logistic probability, team ratings, and game scores—all had predictive accuracy of between 58% and 58.4%, the lowest of the five leagues examined. The regression R^2 values were uniformly low, between 2.4% and 3.6%, while the regression errors were all around 4.20. The modeling approaches did not have strong predictive power for baseball, though each still performed substantially better than chance.

CHAPTER 11

Statistics in Baseball*

Baseball and statistics have a long and inseparable history. More than a quarter of the pages in the 2015 edition of the *Official Baseball Rules* are devoted to statistical matters. This also means that there is a lot of data available to analysts, and far more than just the totals and averages on the back of a baseball card. We can examine the statistical record beyond the boxscore and perform analyses on a more granular level, looking at combinations and sequences of different plays rather than just evaluating sums.

To illustrate this method, in this chapter we will look at the results of two such studies. We will find out what types of plays contribute to creating runs and which players have contributed the most to their teams' chances of winning ballgames. This is not intended to be an exhaustive or authoritative list of which analyses are most important, nor a comprehensive review of the myriad baseball analytics that have evolved in recent years. Rather, what follows is simply a demonstration of how these expanded data sets may be used so that analysts can then be inspired to devise their own statistical experiments.

These types of analyses require us to consider individual plays, game by game. Thankfully, the good people at Retrosheet, a 501(c)(3) charitable organization continuing in the tradition of the old Project Scoresheet, make this data freely available at www.retrosheet.org. This event data is formatted in a series of text files containing comma-separated values. Each game begins with a series of rows containing information about the game itself (date, start time, umpires, attendance, and so forth), followed by the starting lineups for both sides. This is followed by the play-by-play data, which includes the inning, batter, count, pitch sequence, and a coded string representing the play's scoring and the disposition of any baserunners, interspersed with any offensive or defensive substitutions. The

---

* The information used here was obtained free of charge from and is copyrighted by Retrosheet. Interested parties may contact Retrosheet at 20 Sunset Rd., Newark, DE 19711.

Optimal Sports Math, Statistics, and Fantasy.
DOI: http://dx.doi.org/10.1016/B978-0-12-805163-4.00011-6

game ends with a list of all pitchers and the number of earned runs charged to each. A typical game may have around 165 rows in the file.

To make analyses easier to perform, Retrosheet offers a DOS executable called bevent.exe that scans the event files and outputs the data in a more comprehensive format. This utility keeps track of the game conditions as the data progresses and can output up to 97 different values, such as what the home team's score was or who the pitcher and baserunners were at any point in the ballgame. These values can then be loaded into a database system, which can then be employed to perform any manner of analysis.

Such was the case with the two analyses presented here. Retrosheet data was run through the bevent.exe utility and then imported into a PostgreSQL database. We used all data that was available beginning with the 1940 season through the 2015 season, comprising 10,800,256 events involving 11,843 players over 190,049 games.

11.1 RUN CREATION

To score a run in baseball, two things need to happen: the batter must reach base, and then he must find his way home. Often that is accomplished through the actions of a subsequent batter, though it can also occur through the actions of the defense, the pitcher, or himself.

Scoring Runs as a Function of How That Player Reached Base

Table 11.1 lists the different types of events that can take place with respect to a batter during a plate appearance. Reached on Error includes only those errors that allowed the batter to reach base, and not those that may have been committed as part of the same play but after the batter reached base. So if a batter reaches on a base hit and takes second when the right fielder bobbles the ball, that is represented under Single and not Reached on Error. Number of Plays reflects how many instances of that event there were over the span of 76 seasons between 1940 and 2015 as reflected in the available Retrosheet data.

The column Batter Reached Base Safely indicates how many times the batter not only reached base but also did not make an out on the basepaths before the play had concluded. This is why the number of batters who drew a walk is not equal to the figure under the Batter Reached

Table 11.1 Reaching Base by Play Type, 1940–2015

Play Type	Runs Scored			Number of Plays	Runs Per Play	Batter Reached Safely	Runs Per Time on Base
	by Batter	by PR	Total				
Home Run	238,542	0	238,542	238,542	1.0000	238,542	1.0000
Triple	35,917	289	36,206	61,083	0.5927	60,630	0.5972
Double	174,777	2954	177,731	426,075	0.4171	421,983	0.4212
Error	33,174	660	33,834	121,761	0.2779	118,686	0.2851
Single	428,395	8427	436,822	1,680,729	0.2599	1,664,604	0.2624
Hit by Pitch	16,784	759	17,543	67,744	0.2590	67,744	0.2590
Walk	199,888	4263	204,151	821,899	0.2484	821,885	0.2484
Interference	287	3	290	1178	0.2462	1178	0.2462
Fielder's Choice	6431	102	6533	31,976	0.2043	31,502	0.2074
Intentional Walk	12,081	338	12,419	80,094	0.1551	80,094	0.1551
Force Out	28,764	420	29,184	279,066	0.1046	278,987	0.1046
Sacrifice	3562	54	3616	109,431	0.0330	8948	0.4041
Sacrifice Fly	175	2	177	68,007	0.0026	473	0.3742
Strikeout	1053	26	1079	1,566,190	0.0007	4572	0.2360
Double Play	72	1	73	246,285	0.0003	1196	0.0610
Generic Out	5	0	5	4,643,255	0.0000	38	0.1316
Triple Play	0	0	0	293	0.0000	0	—
Total	1,179,907	18,298	1,198,205	10,443,608	0.1147	3,801,062	0.3152

Base Safely column. Since 1940 the available data includes 14 instances in which a batter reached first base on a walk and immediately attempted to advance for whatever reason, only to be tagged out. Most recently this occurred in a 2009 game between the Boston Red Sox and the Atlanta Braves. Ball four to Boston's Kevin Youkilis got by Braves catcher David Ross; Youkilis tried to take second and Ross threw him out. Thirteen of these events are represented in the Walk row; the fourteenth involved Bill Wilson of the Philadelphia Phillies who in 1969 somehow managed to walk into a double play against the Pittsburgh Pirates.

This analysis was done by taking the *events* table in the database (aliased as *e1*) and left-joining it to itself twice (aliased as *pr* and *e2*) such that the three tables had the same game and inning identifiers. A separate join constraint on the *pr* table accounts for any batters who reached base and were removed for pinch runners, while the join condition on *e2* restricts it to only those events in which a run was scored by the batter in *e1* or his pinch runner. (The possibility of a pinch runner for a pinch runner was ignored.) To prevent miscalculations that may arise when teams bat around, two additional constraints were used on each table join. The event number in *e1* must be less than that in *pr* and less than or equal to that in *e2*; the "or equal to" accounts for home runs and errors that allow the batter to score directly. Additionally, the batting team's score in *e1* must be no more than four runs less than the scores in *pr* and *e2*. This query took just under 15 minutes to complete on a typical desktop computer running PostgreSQL 9.4.5 on Microsoft Windows 7.

It's always a good idea to check over the data to see if anything is out of order. Every batter who hits a home run both reaches base safely and scores. For singles, doubles, and triples, a batter could be thrown out trying to take an extra base, so Number of Plays and Batter Reached Base Safely need not be equal. It is possible, but rare, to reach base on a double play, as the two outs could be made on other baserunners.

One figure that jumps out is the 38 batters who reached base via a generic out. This category includes ordinary fly outs, pop outs, line outs, and ground outs that did not get rolled up into one of the other categories. All of the other out types that could allow the batter to reach base (when coupled with a wild pitch, passed ball, fielding error, or a play on another runner) are represented; this total should be 0, not 38. Once the 38 records are identified, we find that they are consistent with ground-ball force outs—an assist for an infielder, a putout for the shortstop or second base-man, runner from first out at second, batter reaches first—yet they are clas-sified as outs in the air. Nearly a third of these suspect records come from two three-game series played in Chicago in 1988, one between the Phillies and the Cubs in June and the other between the Brewers and the White Sox in September. Put together, this evidence points to an inconvenient truth: there are likely problems with the source data.

This is an instructive reminder that as analysts, we need to look into anything that doesn't seem quite right, and find out why our results are what they are. In the case of the 14 walks that didn't reach base, we find a few very rare and unusual plays. The 38 stray generic outs reveal a problem with our source data. And there's always that third and most common possibility, that we've made an error in our analysis.

When batters reached base via a hit, they stayed safe 99.15% of the time, 99.05% excluding home runs. With a walk or when hit by a pitch, they stayed safe 99.999% of the time. But even when their on-base per-centage would not increase—reaching on an error, fielder's choice, force out, dropped third strike, or catcher's interference—hitters still managed to reach base safely 6.30% of the time.

Overall, 31.2% of batters who reached base (or their pinch runners) came around to score. Extra-base hits exceeded the average—home runs 100%, triples 59.2%, doubles 41.4%—but so did both sacrifices (39.9%) and sacrifice flies (37.5%). Of course it's rare to reach base by either of those methods (8.2% for sacrifices and 0.7% for sacrifice flies). Events that ordinarily leave the batter on first saw runners score at generally the same rates: 25.9% for singles, 25.1% for hit batsmen, 24.4% for interference and walks, and 23.2% for dropped third strikes. Though seemingly simi-lar, hit batsmen and walks should be considered separately: when a batter is hit by a pitch the ball is dead, but when he draws a base on balls, the ball is live and there is a chance, albeit very small, that one or more run-ners could either advance another base or make an out. The percentages for fielder's choices (20.6%) and force outs (10.4%) should be, and in fact are, lower than for other types of plays that do not register an out. If a batter reaches on a fielder's choice or a force out, the next batter rarely comes to the plate with no outs; additionally, the fielder's choice and force out numbers include those that ended the inning. Batters given intentional walks scored 15.1% of the time. Intentional walks are not ran-dom events and are only used when the team in the field considers it to be to their advantage, primarily either to bypass a stronger hitter to pitch to a weaker hitter or when first base is unoccupied to set up a force.

Bringing Runners Home

Once there's a runner on base, it's time to bring him home. A batter can do this himself by hitting a home run, but barring that he'll generally need someone else to drive him in.

The types of plays that can bring a runner home are more numerous than those that can put a man on base. A runner can score on a balk or a wild pitch or even a stolen base. There are two event types, the home run and the sacrifice fly, whose very definition includes driving in a run, and for which we should expect that the average number of runs brought home should be at least 1.

This query used for this analysis was much simpler, consisting of only one table join. The *events* table was again joined to itself with aliases *e1* and *e2* with the same constraints on game and inning identifiers and considerations for teams batting around. Unlike the run creation analysis, here we're not concerned with whether the run was scored by the batter or by a pinch runner, simply that the run scored and whether the batter earned an RBI.

Not every run that scores has a corresponding RBI, so in Table 11.2, runs scored and RBIs credited are listed separately.

As with the run-scoring analysis, any results that don't look quite right should be further investigated in case they signal a flaw in our methodology. There were 318 RBIs credited on double plays and one on a triple play. While the *Official Baseball Rules* does hold in Rule 10.04(b)(1) that "[t]he official scorer shall not credit a run batted in when the batter grounds into a force double play or a reverse-force double play," a batter may receive an RBI in other types of double plays, such as one in which the batter hits a sacrifice fly and an out is made on a trailing baserunner. The RBI on a triple play went to Brooks Robinson on September 10, 1964, facing Jim Hannan of the Washington Senators with the bases loaded in the top of the fifth inning. Robinson hit into what started as a conventional 6−4−3 double play that scored Jerry Adair and moved Luis Aparicio from second to third, only for Aparicio to make the third out at home. The one run that scored via defensive indifference took place on June 21, 1989, when the Reds' Lenny Harris scored from third following a pickoff throw to first by the Braves' Jim Acker.

Table 11.2 Play Types that Score Runs, 1940—2015					
Play Type	Number of Plays	Runs Scored	RBI	Runs Per Play	RBI Per Play
Home Run	238,542	379,453	379,453	1.5907	1.5907
Sacrifice Fly	69,019	69,295	69,056	1.0040	1.0005
Triple	61,083	39,346	38,594	0.6441	0.6318
Double	426,075	180,251	178,376	0.4230	0.4186
Single	1,680,729	384,048	373,063	0.2285	0.2220
Error	121,761	24,655	5325	0.2025	0.0437
Wild Pitch	73,944	13,051	–	0.1765	–
Balk	11,579	1977	–	0.1707	–
Fielder's Choice	31,976	5000	3499	0.1564	0.1094
Passed Ball	19,755	3033	–	0.1535	–
Force Out	279,066	20,191	17,747	0.0724	0.0636
Pickoff	31,794	1365	–	0.0429	–
Sacrifice	109,580	4591	3213	0.0419	0.0293
Other Advance	3615	138	–	0.0382	–
Double Play	245,125	8051	318	0.0328	0.0013
Hit by Pitch	67,744	1966	1966	0.0290	0.0290
Interference	1178	32	28	0.0272	0.0238
Stolen Base	144,399	3282	–	0.0227	–
Walk	821,899	17,438	16,843	0.0212	0.0205
Triple Play	282	4	1	0.0142	0.0035
Generic Out	4,643,255	39,988	39,919	0.0086	0.0086
Caught Stealing	61,538	192	–	0.0031	–
Strikeout	1,566,190	871	–	0.0006	–
Defensive Indifference	5559	1	–	0.0002	–
Intentional Walk	80,094	9	3	0.0001	0.0000
Total	10,795,781	1,198,228	1,127,404	0.1110	0.1044

Comparing Estimated and Actual Values

So we have these percentages—the rate of how often a batter who reached base by getting hit by a pitch came around to score, the average number of runs brought home by a double, and so on. We can then use these as coefficients to project run and RBI totals. We'll ignore the possibility of an intentional walk with the bases loaded:

$$Runs \approx 0.2599 \cdot H_{1B} + 0.4171 \cdot H_{2B} + 0.5927 \cdot H_{3B} + HR$$
$$+ 0.2484 \cdot BB + 0.2590 \cdot HBP + 0.1551 \cdot IBB + 0.0330 \cdot SH$$
$$+ 0.0026 \cdot SF + 0.0007 \cdot K + 0.2779 \cdot ReachOnErr$$
$$+ 0.2462 \cdot ReachOnInterference + 0.2043 \cdot FC + 0.1046 \cdot FO$$

$$RBI \approx 0.2285 \cdot H_{1B} + 0.4186 \cdot H_{2B} + 0.6318 \cdot H_{3B} + 1.5907 \cdot HR$$
$$+ 0.0205 \cdot BB + 0.0290 \cdot HBP + 0.0293 \cdot SH + 1.0005 \cdot SF$$
$$+ 0.0437 \cdot ReachOnErr + 0.0238 \cdot ReachOnInterference$$
$$+ 0.1094 \cdot FC + .0636 \cdot FO$$

It may not be practical to project run and RBI totals using all of these parameters, as the values may not be readily available. Websites listing players' batting statistics are not likely to include how many times Curtis Granderson has grounded into a fielder's choice, for example. To work around this, we can focus on the more common "back of the baseball card" stats and roll up the more specific types of outs into a single figure. After combining errors, fielder's choices, force outs, and interference with generic outs (leaving sacrifices and sacrifice flies alone), we arrive at averages of 0.01313 batter-runs and 0.00959 RBI per out. For RBI, we can calculate outs simply as $AB - H$ as sacrifices, sacrifice flies, and catcher's interference do not charge a time at bat. Since a player can reach base on a strikeout, and a strikeout is required to reach base on a wild pitch or passed ball, we'll keep that in the runs formula. We do not need to include strikeouts in the RBI formula because (1) the run scores on the wild pitch or passed ball and not the strikeout itself, and (2) no RBI is awarded for a run that scores on a dropped third strike.

$$Runs \approx 0.2599 \cdot H_{1B} + 0.4171 \cdot H_{2B} + 0.5927 \cdot H_{3B} + HR$$
$$+ 0.2484 \cdot BB + 0.2590 \cdot HBP + 0.1551 \cdot IBB + 0.0330 \cdot SH$$
$$+ 0.0026 \cdot SF + 0.0007 \cdot K + 0.01313 \cdot (AB - H - K)$$

$$RBI \approx 0.2285 \cdot H_{1B} + 0.4186 \cdot H_{2B} + 0.6318 \cdot H_{3B} + 1.5907 \cdot HR$$
$$+ 0.0205 \cdot BB + 0.0290 \cdot HBP + 0.0293 \cdot SH + 1.0005 \cdot SF$$
$$+ 0.00959 \cdot (AB - H)$$

Regressions of the simplified formulas against the actual run and RBI totals had virtually equal R^2 values compared with regressions using the more complicated formulas. For runs, the R^2 dropped from 96.85% to 96.72% and the t-Stat from 681.7 to 668.1, with the standard error rising from 5.19 to 5.50. Projected run totals changed by an average of 0.39 runs, with a maximum of 3.5. For RBI the differences were even smaller, 96.83% to 96.75% for R^2, 5.09 to 5.42 for the standard error, and 679.7 to 671.3 for the t-Stat. The average difference between each formula's projected RBI

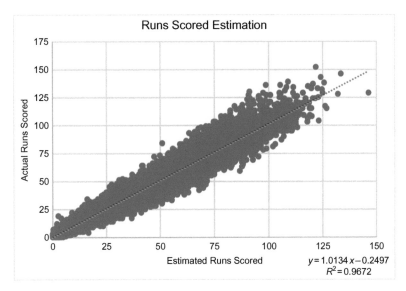

Figure 11.1 Runs Scored Estimation.

Table 11.3 Runs Scored Estimation: 2000−15 MLB		
Statistic	**Value**	**t-Stat**
b_0 (y-Intercept of best-fit line)	−0.250	−4.4
b_1 (Slope of best-fit line)	1.013	668.1
R^2	96.72%	
F-value	446,373.61	
Standard Error	5.499	

total was 0.84 RBI, with a maximum of 3.99. As the changes were rather insignificant, we will use the simpler formulas (Fig. 11.1 and Table 11.3).

The greatest underprojection was 91.2 runs scored for Rafael Furcal in 2003, when his actual total of 130 was third best in the National League. The greatest overprojection was 82.1 runs for David Ortiz in 2014, who scored only 59 runs and was removed for a pinch runner five times. The differences between projected and actual run totals had a standard deviation of 5.51 runs. Of the 5133 players who scored at least 25 runs in a season between 2000 and 2015, 49.1% of the estimated runs scored were within ± 10% of the actual figures. Barry Bonds rated the highest runs-scored projection of 146.3 for his 2001 season, when he scored 129 runs (Fig. 11.2 and Table 11.4).

Among the sample of players who drove in at least 25 runs in a season between 2000 and 2015, estimated RBI totals were within ± 10% of

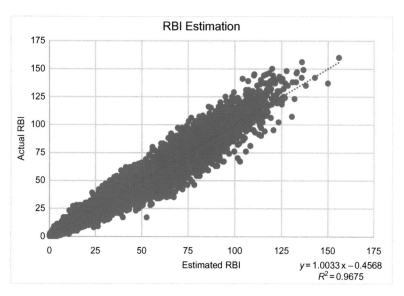

Figure 11.2 RBI Estimation.

Table 11.4 RBI Scored Estimation: 2000–15 MLB		
Statistic	**Value**	**t-Stat**
b_0 (y-Intercept of best-fit line)	−0.457	−8.2
b_1 (Slope of best-fit line)	1.003	671.3
R^2	96.75%	
F-value	450,698.46	
Standard Error	5.418	

actual totals 76.9% of the time, and 10% of the estimates were accurate to within 1 RBI. The closest estimate was 98.998 RBI for Jay Bruce in 2012, when he recorded 99 RBI. The greatest overestimate was for Hanley Ramirez in 2008, when he drove in 67 runs but was expected to drive in 102.7. Preston Wilson of the Colorado Rockies led the National League with 141 RBI in 2003, making his projected total of 101.4 the most substantial underestimate among this sample. The highest estimate was 153.6 RBI for Sammy Sosa in 2001, when he drove in 160 runs.

11.2 WIN PROBABILITY ADDED

Win probability is a measure of how many times in the past a team that found itself in a given game situation went on to win the game. The game situation includes whether the team is the home team or the visiting team, the inning, the number of outs, the number and position of any baserunners, and the score margin. Win probabilities are not probabilities in the strict statistical sense as there is no mathematical theory underlying them; rather they are determined by examining the historical record. Using Retrosheet's play-by-play event data files for the period from 1940 to 2015, comprising over 10.7 million events, we take all of the various permutations of innings, outs, score differentials, baserunners, and play types, correlating that with how often teams in that situation went on to win.

The two teams do not begin the game with equal win probabilities of 0.5, as home teams have won about 54% of the time. The figures fluctuate from play to play based on the changing game situation. By attributing this fluctuation to the pitcher, batter, or baserunner involved in the play, we can measure by how much that player's actions either increased or decreased his club's chance of victory. All events are not created equal; a home run in a tie game in the ninth inning will add far more to the batting team's win probability than would a home run in the third inning with the team up by six runs. We can then aggregate those figures over the course of a game, a season, or a career. This is win probability added, or WPA. Beyond the players, we can also apply WPA to each type of play. This is similar to F.C. Lane's run expectancy analysis, but we can look at much more than just singles, doubles, triples, and home runs, and use more parameters than just the number of outs and positions of any baserunners.

For example, say a team is batting in the bottom of the fourth inning, ahead by one run. There is one out and a runner is on first base. These game conditions occurred 5959 times in the sample for which data was available, and when they did, the batting team won 4302 times. This gives us a win probability of 72.19%. It's the pitcher's turn at bat, and he is asked to sacrifice. If he is successful, the runner will advance to second with two outs, still in the bottom of the fourth, still up by one run. Teams faced these conditions in 3636 games, winning 2570 of them (70.68%). Thus a sacrifice with a runner on first and one out while batting in the home fourth inning with a one-run lead yields a WPA of −1.51 percentage points (72.19−70.68). If the pitcher does not convert the sacrifice and strikes out instead, the runner will stay on first base but with two outs now in the bottom of the fourth ahead by a run; teams in this situation won 4241 of 6072 games, a win probability of 69.85%, and a win probability added for the strikeout of −2.34 percentage points.

Extra innings were treated equally. In other words, a game situation of bases loaded with one out in the top of the 10th inning with a tie score was lumped together with the same conditions but in the top of the 13th inning. One reason for this is simply a matter of sample size; over 157,000 games ended in the ninth inning compared with only 7200 in the 10th and barely 100 past the 16th. The other reason reflects the nature of extra-inning games. Unlike regulation games, which are played for a determined number of innings, extra-inning games persist only as long as the score is still even. When the top of an extra inning begins, the score is always tied, and when the home team takes the lead in the bottom of an extra inning, the game immediately ends. While there may be practical differences between the 10th or 12th or 16th inning, such as the number of substitutes remaining on the bench or in the bullpen, for the purposes of calculating win probability they can be ignored.

Events

Unsurprisingly, home runs added the most to the offensive team's win probability, increasing it by an average of $+13.78$ percentage points. The results also show how the relative power of the home run has declined over the years; the average WPA of a home run was only $+12.84$ percentage points in 2003 compared with $+15.23$ in 1943. In 1968, the so-called Year of the Pitcher, a home run added $+14.56$ percentage points to a team's win probability, the 11th highest figure in the sample. On the other end of the scale, of the 24 seasons with the lowest WPA values for home runs, 21 have taken place in the offensive golden age since the end of the 1994 players' strike.

Triples were right behind at $+9.95$ percentage points, followed by doubles ($+7.26$), reaching on an error ($+4.62$), and singles ($+4.51$). The sacrifice was a minor net negative, lowering a team's win expectancy by an average of -0.29 percentage points. This includes sacrifices employed as a team is playing for a single run in the bottom of the ninth inning, as well as the garden-variety sac bunt by a weak-hitting pitcher early in a ballgame, which for such a pitcher presents a better option than a potential strikeout (-2.62) or double play (-7.49) (Table 11.5).

The offensive team's win probability was hurt the most by triple plays, double plays, baserunners caught stealing, and baserunners out advancing, all plays that add one or more outs and remove one or more baserunners at the same time. A stolen base added $+2.19$ percentage points, or about

Table 11.5 WPA by Play Type, 1940–2015

Play Type	Number of Plays	Total WPA	Average WPA
Home Run	238,542	3,286,679	13.78
Triple	61,083	607,778	9.95
Double	426,075	3,094,461	7.26
Error	121,761	550,197	4.52
Single	1,680,729	7,574,882	4.51
Interference	1178	4150	3.52
Hit by Pitch	67,744	206,530	3.05
Walk	821,899	2,428,502	2.95
Balk	11,579	32,937	2.84
Wild Pitch	73,944	203,722	2.76
Passed Ball	19,755	54,028	2.73
Stolen Base	144,399	316,099	2.19
Intentional Walk	80,094	76,990	0.96
Sacrifice Fly	69,019	54,207	0.79
Defensive Indifference	5559	−89	−0.02
Sacrifice	109,580	−33,016	−0.30
Pickoff	31,794	−72,787	−2.29
Generic Out	4,643,255	−10,818,580	−2.33
Strikeout	1,566,190	−4,145,260	−2.65
Fielder's Choice	31,976	−88,376	−2.76
Force Out	279,066	−875,842	−3.14
Other Advance	3615	−14,695	−4.07
Caught Stealing	61,538	−267,735	−4.35
Double Play	245,125	−1,846,853	−7.53
Triple Play	282	−3899	−13.83

half as much as was lost by an unsuccessful stolen base attempt (−4.35), indicating that in the aggregate, a stolen base attempt would be worth the risk only if it had a probability of success of at least 66.5%. Year to year, this break-even point has fluctuated between 61.4% (1968) and 69.2% (2005). Generally speaking, big league teams have used the stolen base prudently, at least since the expansion era began. From 1940 to 1960, base stealers were successful only 58.1% of the time, far below the break-even level of 64.7% over that period. Since 1961, however, stolen-base success rates have fallen short of the break-even level only four times (1963, 1967, 1991, 1992), and when they did it was by no more than 4.8%. By contrast, more recently the stolen base success rate exceeded the break-even level by at least 9% every season since 2003, peaking at 18.5% in 2007. Part of this may be simply that teams are far more circumspect nowadays with respect to when they will try to steal a base as they were in previous eras, and as a result the running game is a shadow of what it once was. There were 1.15 stolen base attempts per game in 2015, the lowest figure since 1964. Between 1976 and 1999, there were more than 3750 stolen base attempts in every non-strike-shortened season, a level that has not been reached since.

Some plays have identical effects on the game situation, and while their average WPA over time is not exactly equal, any difference is insignificant. For example, a wild pitch (+2.75) and a passed ball (+2.73) both cause at least one baserunner to advance. A walk (+2.95) and a hit batsman (+3.05) both put the batter on first base and advance any runners on base who do not have an open base in front of them. An intentional walk, which also has the same effect, has a much lower WPA (+0.96), but this is because an intentional walk is just that—intentional; the defensive team can call for an intentional walk whenever they feel it would be advantageous to do so, while a regular walk or a hit batsman is largely happenstance. In our sample data covering 1940 through 2015, batters drew 16,843 walks and were hit by pitches 1966 times with the bases loaded, but only three (Lonny Frey in 1941, Barry Bonds in 1998, and Josh Hamilton in 2008) received an intentional walk to drive in a run.

The difference between a runner caught stealing (−4.35) and a runner picked off (−2.28) is much larger. While the two plays are similar, they are not quite the same. The runner has control over when he decides to steal a base and risk being thrown out, while a pickoff is never planned. The CS statistic also includes those runners tagged out after a batter fails to make contact on a hit-and-run or a suicide squeeze. By looking at the data surrounding the circumstances in which runners are caught stealing or picked off, we find that the former has occurred more often when the effect on win probability by trading a baserunner for an out is greater. When a runner was caught stealing, 61.3% of the time he was the only man on base and there were one or two outs, compared with 47.5% of runners picked off.

Batting

For evaluating players, we used only the seasons 1975–2015, as that is as far back as the Retrosheet files go before encountering seasons with incomplete data. Using incomplete data from 1940 to 1974 for compiling play WPA was something we could accept, but for individual players every game really does matter.

When a batter steps to the plate, his team has a certain win probability based on the game conditions. When his turn at bat ends, his team has a second win probability. That second probability could be 0 (he makes the last out of the game) or 1 (he drives in the winning run). The difference between those two numbers represents the win probability added or subtracted during that plate appearance. Over the course of a full season, those numbers add up (or down, as the case may be). In Table 11.6, WPA represents the sum of the number of percentage points by which the win probability of each player's team rose or fell as a result of each of his plate appearances over the course of the entire season.

The lowest single-season WPA during this span was −517.0 for Neifi Pérez of the 2002 Kansas City Royals. That year, Pérez hit .236 with a .260 on-base percentage and .303 slugging percentage for a Royals team that lost 100 games. The next two lowest were Andres Thomas of the 1989 Atlanta Braves (−466.3) and Matt Dominguez of the 2014 Houston Astros (−427.9).

We can normalize the WPA values for individual players by expressing them as a per-plate appearance average. As with any rate statistic, there should be a minimum value for the denominator to minimize statistical insignificance. Without this restriction, the highest single-season WPA average would belong to Dave Sells at 36.32. Sells, a relief pitcher for the Los Angeles Dodgers, came to bat one time during the 1975 season, smacking a go-ahead RBI single in the top of the 11th inning in a September game at the Astrodome. We'll follow the current rule for qualifying for the batting title: at least 3.1 plate appearances per team game,

Table 11.6 Highest WPA, Batting, 1975−2015				
Rank	Batter	Team	Season	WPA
1	Barry Bonds	SF	2004	1309.62
2	Barry Bonds	SF	2001	1198.66
3	Todd Helton	Col	2000	1050.15
4	Mark McGwire	StL	1998	1048.89
5	Barry Bonds	SF	2002	1015.78
6	Albert Pujols	StL	2006	989.99
7	Ryan Howard	Phi	2006	950.58
8	David Ortíz	Bos	2005	940.60
9	David Ortíz	Bos	2006	896.60
10	Jason Giambi	Oak	2005	880.15

Table 11.7 Top 10 Average WPA, Batting, 1975−2015

Rank	Batter	Team	Season	WPA
1	Barry Bonds	SF	2004	2.1226
2	Barry Bonds	SF	2001	1.8052
3	Barry Bonds	SF	2002	1.6598
4	Albert Pujols	StL	2006	1.5615
5	Mark McGwire	StL	1998	1.5402
6	Todd Helton	Col	1998	1.5067
7	Barry Bonds	SF	2003	1.3725
8	Edgar Martínez	Sea	1995	1.3577
9	Ryan Howard	Phi	2006	1.3503
10	David Ortíz	Bos	2005	1.3192

which is 502 for a 162-game schedule. Like the total WPA for hitters, the top 10 averages come from the same era, though the range is now 12 seasons from 1995 to 2006. The list includes each of Barry Bonds' four consecutive National League MVP seasons from 2001 to 2004 (Table 11.7).

Pitching

Every plate appearance begins with one win probability and ends with another. Just as the change can be ascribed to the performance of the hitter in that plate appearance, so it can to the pitcher out on the mound. To be sure, the pitcher can be the statistical victim or beneficiary for the actions of the fielders behind him; e.g., the same ground ball hit with the bases loaded can be either an inning-ending double play or a two-run single depending on the ability and positioning of the second baseman. But as that is the case with many pitching statistics from earned run average (ERA) to batting average against, for the purposes of this particular analysis that is something we can accept.

Dwight Gooden followed up his Rookie of the Year season in 1984 with a career year in 1985: 24−4 with an ERA of 1.53 and 276 strikeouts, taking home the National League Cy Young Award at the tender age of 20. His best game of that season, at least by this measure, was a July 14 start against Houston in the Astrodome; the Mets beat the Astros 1−0 behind Gooden's complete-game five-hitter with 2 walks and 11 strikeouts. Gooden added a total of 76.75 percentage points to the Mets' win probability over the course of the game. Though he may have had other shutouts with more strikeouts and fewer hits allowed, Gooden's

Table 11.8 Top 10 WPA, Pitching, 1975−2015				
Rank	**Pitcher**	**Team**	**Season**	**WPA**
1	Dwight Gooden	NYM	1985	1054.54
2	Orel Hershiser	LAD	1988	802.07
3	Willie Hernández	Det	1984	764.71
4	Greg Maddux	Atl	1995	741.46
5	Zack Greinke	LAD	2015	735.54
6	Roger Clemens	Bos	1990	700.97
7	Jim Palmer	Bal	1975	688.61
8	John Tudor	StL	1985	685.10
9	Frank Tanana	Cal	1976	681.41
10	Vida Blue	Oak	1976	671.71

WPA was higher in this particular game because it remained scoreless until the eighth inning, and consequently each play took on greater magnitude than it would have in a more lopsided contest. The biggest play of the game was when José Cruz, representing the potential winning run with nobody out in the bottom of the ninth, grounded into a 4−6−3 double play, increasing the Mets' win probability from 69.2% to 96.2%. Over the course of the 1985 season, plays like those added up to total of 1054.5 percentage points, 250 more than any other pitcher in a single year since 1975 (Table 11.8).

Pedro Martínez's legendary 1999 season, with his record of 23−4 and 313 strikeouts, ranked 93rd with a total of 459.5. Because of the potential for large swings in win probability late in games, like the 27 percentage points for the Cruz double play, this type of analysis does tend to emphasize the later innings. Gooden had 16 complete games in 1985, while in 1999 Martínez had only 5.

To that end, we can apply the average WPA method for pitchers as well. Given the potential for bias toward the later innings, and the innings in which pitchers pitch over the course of a season are far less uniformly distributed than the innings in which batters come to the plate, the top of the average WPA rankings for pitchers primarily tends toward closers. As with most rate statistics, we should define a minimum value for the denominator. Mariano Rivera averaged 272 batters faced per season while he was the Yankees' full-time closer, so 250 is a good starting point (Tables 11.9−Table 11.12).

Greg Maddux, elected to the Hall of Fame in 2014, had the second-highest career WPA of any pitcher whose career began in 1975 or later.

Table 11.9 Top 10 Average WPA, Pitching, 1975–2015

	Minimum 250 Batters Faced			
Rank	Pitcher	Team	Season	Average WPA
1	Trevor Hoffman	SD	1998	1.8242
2	J.J. Putz	Sea	2007	1.7393
3	Dennis Eckersley	Oak	1990	1.717
4	José Mesa	Cle	1995	1.7084
5	Eric Gagné	LAD	2003	1.7025
6	Mark Melancon	Pit	2015	1.6681
7	Tyler Clippard	Was	2011	1.6176
8	Randy Myers	NYM	1988	1.607
9	Jim Johnson	Bal	2012	1.5495
10	Jay Howell	LAD	1989	1.5128

Table 11.10 Season Batting and Pitching WPA Leaders, 2006–2015

Batter	Team	WPA	Season	Pitcher	Team	WPA
Albert Pujols	StL	990.0	2006	Joe Nathan	Min	385.5
Alex Rodríguez	NYY	739.5	2007	J.J. Putz	Sea	478.3
Manny Ramírez	Bos/LAD	737.7	2008	Cliff Lee	Cle	520.9
Prince Fielder	Mil	786.5	2009	Chris Carpenter	StL	544.4
Miguel Cabrera	Det	823.4	2010	Roy Halladay	Phi	490.5
José Bautista	Tor	799.6	2011	Tyler Clippard	Was	546.8
Joey Votto	Cin	630.9	2012	Felix Hernández	Sea	444.7
Chris Davis	Bal	800.0	2013	Clayton Kershaw	LAD	585.5
Mike Trout	LAA	643.7	2014	Clayton Kershaw	LAD	600.8
Anthony Rizzo	ChC	601.2	2015	Zack Greinke	LAD	735.5

Table 11.11 2015 Batting and Pitching WPA Leaders

Batter	Team	WPA	Rank	Pitcher	Team	WPA
Anthony Rizzo	ChC	601.2	1	Zack Greinke	LAD	735.5
Joey Votto	Cin	577.3	2	Jake Arrieta	ChC	619.1
Paul Goldschmidt	Ari	556.6	3	Clayton Kershaw	LAD	553.1
Josh Donaldson	Tor	539.7	4	Mark Melancon	Pit	500.4
Bryce Harper	Was	507.0	5	Tony Watson	Pit	435.8
Mike Trout	LAA	495.4	6	Dallas Keuchel	Hou	412.1
Matt Carpenter	StL	488.6	7	Wade Davis	KC	387.4
Andrew McCutchen	Pit	447.9	8	Jacob deGrom	NYM	377.6
Chris Davis	Bal	399.6	9	John Lackey	StL	360.4
Carlos González	Col	396.2	10	Max Scherzer	Was	358.7

Table 11.12 Career WPA Leaders, Debuting in 1975 or Later, Through 2015

Batter	Career	WPA	Rank	Pitcher	Career	WPA
Barry Bonds	1986–2007	12,460.2	1	R. Clemens	1984–2007	5275.1
Frank Thomas	1990–2008	6766.3	2	G. Maddux	1986–2008	4802.9
Manny Ramírez	1993–2011	6694.2	3	P. Martínez	1992–2009	3963.4
Alex Rodríguez	1994–	6637.7	4	R. Johnson	1988–2009	3293.9
Albert Pujols	2001–	6437.9	5	C. Kershaw	2008–	3247.6
Jim Thome	1991–2012	6426.8	6	M. Rivera	1995–2013	3065.2
Rickey Henderson	1979–2003	6343.2	7	J. Smoltz	1988–2009	2923.8
Chipper Jones	1993–2012	6134.4	8	F. Hernández	2005–	2656.9
Todd Helton	1997–2013	6015.1	9	R. Halladay	1998–2013	2579.7
Jeff Bagwell	1991–2005	5904.9	10	T. Glavine	1987–2008	2524.2

His achievements on the mound did not translate to success at the plate, however; his career batting WPA of -1608.5 was the fifth lowest among hitters of that same era.

11.3 CONCLUSION

The Retrosheet event data loaded into a database system provides analysts with the tools to perform many types of research in the field of baseball analytics that are not possible using statistics aggregated at the season level. Readers are encouraged to obtain this data and familiarize themselves with it.

CHAPTER 12

Fantasy Sports Models

12.1 INTRODUCTION

In this chapter we apply our different sports modeling techniques to fantasy sports. We show how these models can be used to predict the number of points that a player will score against an opponent and we show how these models can be used by gamers looking to draft or pick a fantasy sports team for a season or weekly contest. These models and approaches can be applied to any type of fantasy sports scoring system as well as used to compute the probability of achieving at least a specified number of points. These models can also be used by professional managers and/or coaches to help determine the best team to field against an opposing team.

The five models that we apply to fantasy sports include:

1. Game Points Model
2. Team Statistics Model
3. Logistic Probability Model
4. Logit Spread Model
5. Logit Points Model

12.2 DATA SETS

We apply our models to fantasy sports competitions using team and player data from the NFL for the 2015−2016 season[1]. We use the first 16 weeks of game data (all data that was available at time the chapter was written). We show how these models can be used to predict player performance for three different positions:

1. Running Backs (RB)
2. Wide Receivers (WR)
3. Quarterbacks (QB)

---

[1] All data used in this chapter was compiled and cross-checked using multiple public sources. The goal of this chapter is to introduce the reader to advanced mathematical and statistical techniques that can be applied to player evaluation and point prediction for fantasy sports competitions.

Optimal Sports Math, Statistics, and Fantasy.
DOI: http://dx.doi.org/10.1016/B978-0-12-805163-4.00015-3

To demonstrate our modeling approach, we selected the player from each NFL team who played in the most games for the team at that position. Therefore, there were 32 players for each position.

The one exception in our analysis was for the wide receiver position, where we included 33 players. The 33rd player in our analysis was Will Tye (tight end) from the NY Giants. Why Will Tye? As a fellow Stony Brook athlete and alumni, we simply decided to include players from our alma mater. Readers can apply the techniques of this chapter to any player of their choice using any specified scoring system[2,3].

A list of players, positions, and games played is shown in Tables 12.1–12.3.

Important Note:

When developing fantasy sports models, it is important to highlight that different fantasy sports competitions have different scoring systems. Depending on the scoring system used, different players may have different performance rankings, expected number of points, and different accuracy measures from the model. For example, in football, scoring systems may provide points for the total yards gained by a player and/or for the total points scored by the player. It may turn out that a model may be very accurate in predicting the number of points that one player will score during a game and not as accurate in predicting the number of yards that player will achieve during a game, and vice versa for a different player. In this chapter we show how our models can be applied to different scoring systems.

12.3 FANTASY SPORTS MODEL

The main difference between sports team prediction models and fantasy sports models is that the fantasy sports models are constructed from the player's perspective. This means that we are treating each player as an individual team and estimating player-specific parameters (similar to team-specific parameters discussed previously). And like the team prediction models, we do need to have on hand a large-enough number of games and degrees of freedom to develop accurate models.

To help resolve the degree of freedom issue and to ensure that we have enough data points and game observations to construct an accurate model, we do not include all of the explanatory factors from the team

[2] In Spring 1987, the Stony Brook men's soccer team beat the Villanova men's soccer team twice. The second victory was in the Spring Soccer Cup Finals. Stony Brook finished the tournament undefeated.

[3] Jim Sadler, QED!

Table 12.1 Running Back (RB)						
Number	Team	Pos	Games	Player	Yards	Points
1	ARI	RB	15	David Johnson	979	72
2	BAL	RB	15	Javorius Allen	799	18
3	CAR	RB	15	Mike Tolbert	346	24
4	CIN	RB	15	Giovani Bernard	1154	12
5	CLE	RB	15	Duke Johnson	868	12
6	DAL	RB	15	Darren McFadden	1272	18
7	DEN	RB	15	Ronnie Hillman	847	36
8	DET	RB	15	Theo Riddick	797	18
9	GNB	RB	15	James Starks	951	30
10	IND	RB	15	Frank Gore	1158	42
11	MIA	RB	15	Lamar Miller	1206	60
12	MIN	RB	15	Adrian Peterson	1639	60
13	NYG	RB	15	Rashad Jennings	983	18
14	OAK	RB	15	Latavius Murray	1242	36
15	PHI	RB	15	Darren Sproles	661	18
16	SDG	RB	15	Danny Woodhead	1029	54
17	TAM	RB	15	Doug Martin	1565	42
18	ATL	RB	14	Devonta Freeman	1540	78
19	CHI	RB	14	Jeremy Langford	802	42
20	HOU	RB	14	Chris Polk	439	12
21	NYJ	RB	14	Chris Ivory	1206	48
22	PIT	RB	14	DeAngelo Williams	1254	66
23	SEA	RB	14	Fred Jackson	351	12
24	WAS	RB	14	Alfred Morris	706	6
25	NOR	RB	13	C.J. Spiller	351	12
26	STL	RB	13	Todd Gurley	1296	60
27	BUF	RB	12	LeSean McCoy	1187	30
28	JAX	RB	12	T.J. Yeldon	1019	18
29	TEN	RB	12	Antonio Andrews	731	24
30	KAN	RB	11	Charcandrick West	816	30
31	NWE	RB	11	LeGarrette Blount	749	42
32	SFO	RB	7	Carlos Hyde	523	18

Note: Data for 2015−2016 Season through Week 16.

prediction models. In these cases, we can simply include only the team's opponent's explanatory factors (e.g., input data). This allows us to develop statistically significant models using a smaller data set, which is important for football where we do not have a large number of games for each team. For other sports (basketball, baseball, soccer, and hockey) readers can experiment with variations of these models following the formulations shown in each of the sports chapters.

The model formulations for football fantasy sports models are shown below for each of the five models. We transform the nonlinear models into linear regression forms for simplicity but the user can also apply the more

Table 12.2 Wide Receiver (WR)

Number	Team	Pos	Games	Player	Yards	Points
1	ARI	WR	15	Larry Fitzgerald	1160	48
2	ATL	WR	15	Julio Jones	1722	48
3	CIN	WR	15	A.J. Green	1263	54
4	DEN	WR	15	Demaryius Thomas	1187	30
5	DET	WR	15	Calvin Johnson	1077	48
6	GNB	WR	15	Randall Cobb	839	36
7	HOU	WR	15	DeAndre Hopkins	1432	66
8	IND	WR	15	T.Y. Hilton	1080	30
9	JAX	WR	15	Allen Robinson	1292	84
10	MIA	WR	15	Jarvis Landry	1201	30
11	NOR	WR	15	Brandin Cooks	1134	54
12	NYG	WR	15	Rueben Randle	718	42
13	NYJ	WR	15	Brandon Marshall	1376	78
14	OAK	WR	15	Michael Crabtree	888	48
15	PHI	WR	15	Jordan Matthews	943	36
16	PIT	WR	15	Antonio Brown	1675	54
17	SEA	WR	15	Doug Baldwin	1023	84
18	STL	WR	15	Tavon Austin	845	54
19	WAS	WR	15	Pierre Garcon	728	30
20	BAL	WR	14	Kamar Aiken	868	30
21	CAR	WR	14	Ted Ginn	799	60
22	CLE	WR	14	Travis Benjamin	925	30
23	DAL	WR	14	Terrance Williams	667	18
24	KAN	WR	14	Jeremy Maclin	1030	42
25	MIN	WR	14	Jarius Wright	450	0
26	BUF	WR	13	Robert Woods	552	18
27	NWE	WR	13	Danny Amendola	678	18
28	SDG	WR	13	Malcom Floyd	561	18
29	SFO	WR	13	Anquan Boldin	718	18
30	TAM	WR	13	Mike Evans	1109	18
31	TEN	WR	13	Harry Douglas	386	12
32	CHI	WR	10	Marquess Wilson	464	6
33	NYG	TE/WR	9	Will Tye	390	12

Note: Data for 2015−16 Season through Week 16.

sophisticated form of the model. Our transformation to a linear form in these situations is mathematically correct since our output variable can take on any real value including both positive and negative values. In our cases, since we are only estimating the total yards per game or total points scored per game per player, our value will only be zero or positive.

The data used as input for these models is from the Chapter 6, Football and is provided in Table 12.4 as a reference.

These transformed fantasy models are as follows:

Table 12.3 Quarterback (QB)

Number	Team	Pos	Games	Player	Yards	Points
1	WAS	QB	15	Kirk Cousins	4038	186
2	TAM	QB	15	Jameis Winston	3915	162
3	SEA	QB	15	Russell Wilson	4368	192
4	SDG	QB	15	Philip Rivers	4587	162
5	OAK	QB	15	Derek Carr	3919	186
6	NYJ	QB	15	Ryan Fitzpatrick	3982	186
7	NYG	QB	15	Eli Manning	4195	198
8	NWE	QB	15	Tom Brady	4726	234
9	MIN	QB	15	Teddy Bridgewater	3322	102
10	MIA	QB	15	Ryan Tannehill	3993	138
11	KAN	QB	15	Alex Smith	3767	120
12	JAX	QB	15	Blake Bortles	4499	222
13	GNB	QB	15	Aaron Rodgers	3862	186
14	DET	QB	15	Matthew Stafford	4107	180
15	CAR	QB	15	Cam Newton	4170	246
16	ATL	QB	15	Matt Ryan	4320	114
17	ARI	QB	15	Carson Palmer	4566	210
18	NOR	QB	14	Drew Brees	4566	192
19	CHI	QB	14	Jay Cutler	3604	120
20	PHI	QB	13	Sam Bradford	3446	102
21	CIN	QB	13	Andy Dalton	3392	168
22	BUF	QB	13	Tyrod Taylor	3374	138
23	TEN	QB	12	Marcus Mariota	3111	132
24	STL	QB	11	Nick Foles	2072	48
25	PIT	QB	11	Ben Roethlisberger	3607	108
26	HOU	QB	10	Brian Hoyer	2401	108
27	BAL	QB	10	Joe Flacco	2814	102
28	CLE	QB	9	Johnny Manziel	1730	42
29	SFO	QB	8	Colin Kaepernick	1871	42
30	IND	QB	8	Matt Hasselbeck	1705	54
31	DEN	QB	8	Peyton Manning	2141	54
32	DAL	QB	8	Matt Cassel	1353	30

Note: Data for 2015–2016 Season through Week 16.

Game Scores Model

$$Y = b_0 + b_1 \cdot HF + b_2 \cdot APA + \varepsilon$$

The variables of this model are:

Y = player's total yards in the game or player's points scored in the game

HF = home field dummy variable, 1 = home game, 0 = away game

APA = away team's average points allowed per game

b_0, b_1, b_2 = player-specific parameters for game scores model

ε = *model error*

Table 12.4 Team Statistics

Team ID	Team Name	Team	Logistic Probability	Logit Spread	Logit Home Points	Logit Away Points	Pi Total Points	Off PPG	Def PPG	Offensive		Defensive	
										OFF Y/R	OFF Y/PA	DEF Y/R	DEF Y/PA
1	Arizona Cardinals	ARI	3.3677	2.8259	2.2333	0.3460	1.2856	29.44	21.22	4.14	8.07	4.04	6.64
2	Atlanta Falcons	ATL	1.1595	1.0353	1.4412	2.1009	1.8477	21.19	21.56	3.81	7.06	4.04	6.92
3	Baltimore Ravens	BAL	0.7351	1.3195	1.3435	1.6784	1.4663	20.50	25.06	3.86	6.32	3.97	6.86
4	Buffalo Bills	BUF	1.2585	1.5760	1.4296	1.4281	1.3819	23.69	22.44	4.78	7.19	4.39	6.60
5	Carolina Panthers	CAR	3.9371	2.7919	2.6214	0.8619	1.8408	36.88	19.42	4.27	7.34	3.91	5.90
6	Chicago Bears	CHI	1.4676	1.4046	1.5640	1.8577	1.6993	20.94	24.81	3.96	7.00	4.47	7.00
7	Cincinnati Bengals	CIN	2.7765	2.8496	2.2185	0.3975	1.2208	27.19	17.47	3.86	7.57	4.41	6.18
8	Cleveland Browns	CLE	0.0590	0.7858	1.1023	2.1911	1.6415	17.38	27.00	4.02	6.21	4.49	7.85
9	Dallas Cowboys	DAL	0.2256	0.7139	1.6343	2.9357	2.5809	17.19	23.38	4.63	6.56	4.20	7.19
10	Denver Broncos	DEN	3.4468	2.4491	2.0696	0.7459	1.3536	26.38	17.89	4.02	6.46	3.36	5.81
11	Detroit Lions	DET	1.7342	1.5190	1.8074	1.8729	1.9491	22.38	25.00	3.77	6.66	4.22	7.18
12	Green Bay Packers	GNB	2.6716	2.2653	1.8926	0.8111	1.2819	26.44	20.39	4.34	6.09	4.44	6.77
13	Houston Texans	HOU	1.4241	1.3096	0.8040	1.1229	0.7994	21.19	20.18	3.71	6.08	4.09	6.18
14	Indianapolis Colts	IND	1.2387	0.6632	0.4748	1.6894	0.9150	20.81	25.50	3.63	5.98	4.32	7.03
15	Jacksonville Jaguars	JAX	0.1725	0.6406	0.5410	1.8561	0.9814	23.50	28.00	4.16	6.77	3.68	7.13
16	Kansas City Chiefs	KAN	2.6741	2.8569	2.2107	0.3767	1.1296	28.88	17.06	4.56	6.98	4.14	5.90
17	Miami Dolphins	MIA	0.5207	0.7122	0.9434	2.1122	1.4678	19.38	24.31	4.35	6.48	4.01	7.38
18	Minnesota Vikings	MIN	2.6537	2.2781	2.2434	1.2107	1.7490	23.38	18.35	4.51	6.43	4.20	6.65
19	New England Patriots	NWE	2.5184	2.5070	2.2705	0.9118	1.5782	31.44	20.11	3.66	7.00	3.89	6.46
20	New Orleans Saints	NOR	0.9549	0.6943	1.7791	3.0750	2.9816	25.38	29.75	3.76	7.45	4.91	8.35
21	New York Giants	NYG	0.7038	1.0749	1.2922	2.0010	1.5849	26.25	27.63	3.99	6.98	4.37	7.50
22	New York Jets	NYJ	1.5854	1.7343	1.8037	1.5552	1.7389	24.19	19.63	4.17	6.72	3.58	6.26
23	Oakland Raiders	OAK	1.5202	1.5912	1.6525	1.6111	1.6271	22.44	24.94	3.94	6.40	4.13	6.46
24	Philadelphia Eagles	PHI	1.0017	0.9314	1.1111	2.0042	1.4896	23.56	26.88	3.93	6.56	4.50	6.75
25	Pittsburgh Steelers	PIT	2.5387	2.6647	2.5285	0.9985	1.8817	28.56	19.89	4.53	7.82	3.78	6.81
26	San Diego Chargers	SDG	0.5232	1.2922	1.4098	1.8552	1.6703	20.00	24.88	3.46	6.88	4.81	7.42
27	San Francisco 49ers	SFO	1.0602	0.7764	0.3741	1.4217	0.5985	14.88	24.19	3.96	6.30	4.01	7.61
28	Seattle Seahawks	SEA	2.7112	2.9311	2.7201	0.7846	1.9110	28.56	17.61	4.52	7.63	3.49	6.18
29	St. Louis Rams	STL	1.6308	1.4755	1.5268	1.6459	1.6022	17.50	20.63	4.56	5.93	4.02	6.81
30	Tampa Bay Buccaneers	TAM	0.6048	0.5981	1.0367	2.3792	1.7499	21.38	26.06	4.76	7.20	3.45	7.10
31	Tennessee Titans	TEN	−0.4483	0.2571	0.6038	2.4502	1.5342	18.69	26.44	4.00	6.36	3.89	7.31
32	Washington Redskins	WAS	1.3597	1.2521	1.2822	1.6995	1.4330	25.38	24.35	3.69	7.36	4.80	7.21
33	HF		0.2124	0.2232	0.0336	0.0116	0.0272	#N/A	#N/A	#N/A	#N/A	#N/A	#N/A

Team Statistics Model

$$Y = b_0 + b_1 \cdot HF + b_2 \cdot HTDYR + b_3 \cdot ATDYPA + \varepsilon$$

The variables of the model are:

Y = player's total yards in the game or player's points scored in the game

HF = home field dummy variable, 1 = home game, 0 = away game

$ATDYR$ = away team average yards per rush allowed

$ATDYPA$ = away team average yards per pass attempt allowed

b_0, b_1, b_2, b_3 = model parameter, sensitivity to the variable

ε = *model error*

Logistic Probability Model

$$Y = b_0 + b_1 \cdot HF + b_2 \cdot Away\ Rating + \varepsilon$$

The variables of the model are:

Y = player's total yards in the game or player's points scored in the game

HF = home field dummy variable, 1 = home game, 0 = away game

Away Rating = away team's logistic probability rating

b_0, b_1, b_2 = model parameter, sensitivity to the variable

ε = *model error*

Logit Spread Model

$$Y = b_0 + b_1 \cdot HF + b_2 \cdot Away\ Logit\ Spread + \varepsilon$$

The variables of the model are:

Y = player's total yards in the game or player's points scored in the game

HF = home field dummy variable, 1 = home game, 0 = away game

Away Logit Spread = away team's logit spread rating

b_0, b_1, b_2 = model parameter, sensitivity to the variable

ε = *model error*

Logit Points Model

$$Y = b_0 + b_1 \cdot HF + b_2 \cdot Logit\ Points + \epsilon$$

The variables of the model are:

Y = player's total yards in the game or player's points scored in the game

HF = home field dummy variable, 1 = home game, 0 = away game

Logit Points = opponent's away logit points rating if a home game and opponent's home logit points rating if an away game

b_0, b_1, b_2 = model parameter, sensitivity to the variable

ε = *model error*

Predicting Points

Similar to the approaches presented in the chapters above, these regression equations and player-specific parameters can be used to predict the number of points a player will score for whatever scoring system is being used by the fantasy sports competition. Readers can refer to the specific sports chapters in this book for further insight into the prediction model.

Points Probability

We can also use the regression results from these models to estimate the probability that the player will score more than a specified number of points or less than a specified number of points, as well as the probability that they will score within an interval of points such as between 250 and 300 yards in a game. These techniques are described in detail in the chapter for each sport for the specified model.

12.4 REGRESSION RESULTS

The results for our five fantasy sports models for Will Tye are shown in Table 12.5A and B. The models used data for the 2015−16 season for the first 16 weeks of the season.

Table 12.5A shows the regression results for predicting total yards per game. The model performance results measured from the model's R^2 goodness of fit ranged from a low of $R^2 = 21.4\%$ for the logistic probability model to a high of $R^2 = 41.6\%$ for the team statistics model.

Table 12.5B shows the regression results for predicting total points per game. The model performance results ranged from a low of $R^2 = 5.9$ for the team statistics model to a high of $R^2 = 34.8$ for the logit points model.

Table 12.5A Will Tye Fantasy Sports Model: Total Yards Per Game Model

Games Points Model

Est	OppPAG	HF	Const
Beta	−4.500	−2.542	150.561
Se	4.430	24.622	112.955
R^2/F	0.348	19.870	#N/A
SeY/df	1.337	5.000	#N/A
SSE/TSS	1055.490	1974.010	#N/A
t-Stat	−1.016	−0.103	1.333

Team Statistics Model

Est	Opp/YPA	OppY/R	HF	Const
Beta	0.409	22.931	26.125	−62.774
Se	15.481	19.579	18.183	132.699
R^2/F	0.416	21.030	#N/A	#N/A
SeY/df	0.950	4.000	#N/A	#N/A
SSE/TSS	1260.414	1769.086	#N/A	#N/A
t-Stat	0.026	1.171	1.437	−0.473

Logistic Probability Model

Est	Team Rating	HF	Const
Beta	0.349	17.510	35.945
Se	9.549	20.438	13.723
R^2/F	0.214	21.821	#N/A
SeY/df	0.681	5.000	#N/A
SSE/TSS	648.637	2380.863	#N/A
t-Stat	0.037	0.857	2.619

Logit Spread Model

Est	Logit Spread	HF	Const
Beta	5.450	12.120	31.490
Se	13.078	20.718	15.670
R^2/F	0.240	21.455	#N/A
SeY/df	0.791	5.000	#N/A
SSE/TSS	727.943	2301.557	#N/A
t-Stat	0.417	0.585	2.010

Logit Points Model

Est	Logit Points	HF	Const
Beta	37.575	14.457	−4.833
Se	28.559	13.570	32.611
R^2/F	0.416	18.810	#N/A
SeY/df	1.781	5.000	#N/A
SSE/TSS	1260.453	1769.047	#N/A
t-Stat	1.316	1.065	−0.148

Note: Data for 2015–16 through Week 16.

Table 12.5B Will Tye Fantasy Sports Model: Total Points Per Game Model

Game Points Model

Est	OppPAG	HFA	Const
Beta	−0.747	−3.408	20.462
Se	0.652	3.625	16.630
R^2/F	**0.208**	**2.925**	#N/A
SeY/df	0.655	5.000	#N/A
SSE/TSS	11.213	42.787	#N/A

Logistic Probability Model

Est	Logistic Probability	HFA	Const
Beta	1.506	−2.114	0.187
Se	1.271	2.720	1.826
R^2/F	**0.219**	**2.904**	#N/A
SeY/df	0.702	5.000	#N/A
SSE/TSS	11.844	42.156	#N/A

Logit Points Model

Est	Logit Points	HFA	Const
Beta	−6.580	0.621	8.695
Se	4.030	1.915	4.601
R^2/F	**0.348**	**2.654**	#N/A
SeY/df	1.333	5.000	#N/A
SSE/TSS	18.783	35.217	#N/A

Team Statistics Model

Est	Opp/Y/PA	OppY/R	HFA	Const
Beta	−1.231	−0.466	−0.838	12.206
Se	2.624	3.318	3.081	22.489
R^2/F	**0.059**	**3.564**	#N/A	#N/A
SeY/df	0.084	4.000	#N/A	#N/A
SSE/TSS	3.191	50.809	#N/A	#N/A

Logit Spread Model

Est	Logit Spread	HFA	Const
Beta	1.512	−1.631	0.179
Se	1.886	2.987	2.259
R^2/F	**0.114**	**3.093**	#N/A
SeY/df	0.321	5.000	#N/A
SSE/TSS	6.152	47.848	#N/A

Game Points	0.208
Team Stats	**0.059**
Logistic Probability	0.219
Logit Spread	0.114
Logit Points	**0.348**

Note: Data: 2015—16 through Week 16.

Analysis of these results shows that for almost all models and for all predicted output values, the model parameters have a very low t-statistic, indicating that the explanatory variable may not be significant. However, the F-Stat of these models shows that the combination of explanatory factors is significant (i.e., all the input variables are not statistically equal to zero).

One of the reasons for the low t-statistics is a small number of observations. For example, we have nine observations for Will Tye for our analysis period (first 16 weeks of games) and we are estimating three or four model parameters. The small degrees of freedom for these models is the cause of the low t-statistic but high R^2. The small number of observations per player is what makes predicting player performance for fantasy sports more difficult that predicting team performance, where we have at least the same number of game observations (and often more data points if the player did not participate in all games), and we also have observations across common opponents, which results in a higher level of statistical significance.

There are two ways to increase the statistical accuracy of fantasy sports models. First, we can utilize a larger data sample, which is not always possible since we are limited to the number of games played and we often want to predict results before the end of the season. A second way to increase statistical accuracy is to perhaps model each half of each game, thus doubling the data sample size. Depending on the number of times the team has possession in each quarter, it might also be possible to model player performance in each quarter as well, resulting in an even larger data set—almost four times as many data points.

Example

The NY Giants (home) are playing the NY Jets at home. We are interested in determining the total number of yards for Will Tye using the logistic probability model with parameters shown in Table 12.5A and the NY Jets' logistic probability rating (Table 12.4) as follows:

$$Est.\,Total\ Yards = 35.945 + 14.457 + 0.349 \cdot 1.5854 = 54.008$$

If we are interested in the probability that Will Tye will have at least 45 yards in the game, we use the standard error of the regression along with the expected number of yards as follows:

$$Prob = 1 - NormCDF(45, 54, 9.549) = 83\%$$

12.5 MODEL RESULTS

Fantasy sports model parameters for all of the selected players for both predicted yards per game and predicted points per game are provided in Tables 12.6−12.11. These parameters can be used to estimate the points from any player against any team.

Table 12.6 shows parameters for predicted yards per game for running backs. Table 12.7 shows parameters for predicted total points per game for running backs. Table 12.8 shows parameters for predicted yards per game for wide receivers. Table 12.9 shows parameters for predicted total points per game for wide receivers. Table 12.10 shows parameters for total yards per game for quarterbacks. Table 12.11 shows parameters for total points per game for quarterbacks.

Overall, we found that the models performed slightly better for predicting total points per game compared to total yards per game by player. This is primarily due to the often large variation in player yards per game across games. Total points per game by player had a lower variation from game to game.

Additionally, we found that the team statistics models proved to be the best-performing models based on our R^2 goodness of fit measure for both yards per game and points per game. This is most notable since this model was found to have the lowest predictive power of team performance across all our models. However, it was the best-performing model of player performance.

Readers interested in expanding and improving these results for fantasy sports could use variations of the team statistics models by incorporating a larger number of team statistics input variables and using a larger data set by modeling each game by half or by quarter. This could increase the total data set by two to four times, thus providing better results and a higher level of statistical accuracy.

12.6 CONCLUSION

In this chapter we applied five of our sports models to fantasy sports. We showed how the model and data from the team models can be applied to fantasy sports by player using a linear regression modeling approach. These models can be easily applied for a variety of different fantasy sports scoring systems such as for total yards per game or total points per game for football. These techniques can be easily applied to different sports and

Table 12.6 Running Back Prediction Model: Total Yards Per Game

Number	Player	Game Points Model				Team Statistics Model					Logistic Probability Model				Logit Spread Model				Logit Points Model			
		R^2	Const	HF	OppPAG	R^2	Const	HF	OppY/R	OppY/PA	R^2	Const	HF	Logistic Probability	R^2	Const	HF	Logit Spread	R^2	Const	HF	Logit Points
12	Adrian Peterson	0.232	-65.227	-16.008	8.211	0.123	-48.992	-34.050	39.186	1.866	0.237	177.050	-28.232	-28.013	0.242	181.939	-15.785	-36.165	0.257	52.293	-14.733	43.793
24	Alfred Morris	0.396	-13.483	46.113	1.472	0.504	-135.603	43.171	34.301	2.346	0.404	3.424	56.585	11.148	0.394	5.964	54.866	9.048	0.420	45.968	57.741	-15.509
29	Antonio Andrews	0.768	-17.982	-37.708	4.266	0.714	-79.541	-35.290	9.045	18.246	0.704	100.114	-35.897	-13.309	0.742	109.900	-38.625	-19.969	0.463	78.952	-37.370	2.395
25	C.J. Spiller	0.045	10.399	11.241	0.440	0.188	-146.212	9.452	16.223	14.826	0.105	29.806	8.845	-5.628	0.098	31.577	10.459	-8.542	0.222	-1.483	-4.335	18.267
32	Carlos Hyde	0.001	80.059	2.212	-0.308	0.074	-155.844	9.933	29.262	15.740	0.007	81.853	2.958	-4.018	0.044	105.602	3.149	-14.897	0.001	68.100	5.178	2.428
30	Charcandrick West	0.080	137.558	23.325	-3.199	0.739	223.704	7.804	-125.124	55.506	0.175	55.006	22.874	17.584	0.082	34.953	20.466	17.342	0.113	131.227	10.214	-34.136
21	Chris Ivory	0.244	130.262	51.139	-2.625	0.430	-316.389	37.154	70.186	13.573	0.240	59.339	51.705	7.437	0.229	66.376	47.136	1.326	0.307	105.342	65.691	-28.814
20	Chris Polk	0.116	61.509	-7.587	-1.163	0.139	63.830	-6.728	7.006	-8.535	0.109	29.515	-8.158	3.561	0.156	26.282	-8.076	6.040	0.063	36.548	-8.870	-1.017
16	Danny Woodhead	0.264	-30.226	14.698	3.992	0.536	-311.595	19.203	51.176	23.326	0.211	77.101	18.502	-11.949	0.190	83.742	16.698	-14.697	0.332	28.538	-7.632	-18.789
6	Darren McFadden	0.083	177.158	-23.063	-3.446	0.114	283.229	-28.457	-19.099	-15.245	0.014	88.212	-10.479	0.578	0.055	75.419	-19.778	12.495	0.056	113.920	-21.061	2.418
15	Darren Sproles	0.158	116.625	-11.704	-2.909	0.207	135.210	-20.775	-31.037	6.633	0.106	51.022	-19.856	1.026	0.106	51.590	-20.203	0.768	0.108	48.414	4.831	-17.763
1	David Johnson	0.020	12.805	-2.966	2.148	0.224	-51.588	-14.636	93.698	-38.705	0.022	78.518	-3.282	-10.337	0.069	93.499	-4.328	-19.697	0.054	91.026	-4.776	34.627
22	DeAngelo Williams	0.098	72.487	41.651	-0.475	0.138	162.207	48.735	15.294	-24.965	0.148	31.455	48.737	14.825	0.126	29.306	52.068	14.603	0.176	-8.144	63.131	-3.609
18	Devonta Freeman	0.006	143.780	1.702	-1.305	0.073	-76.239	9.601	9.506	20.397	0.088	124.031	6.426	-15.151	0.038	125.688	3.147	-16.232	0.003	114.834	4.987	-17.470
17	Doug Martin	0.052	72.266	-21.395	1.878	0.160	113.645	2.122	65.234	-40.072	0.071	104.944	-16.456	10.080	0.069	100.698	-19.077	16.045	0.074	138.584	-2.304	-12.264
5	Duke Johnson	0.159	5.423	-17.102	2.880	0.284	-116.360	-13.182	5.318	24.106	0.091	73.817	-12.172	-4.857	0.127	83.810	-15.019	-9.159	0.128	88.883	-20.728	-3.239
10	Frank Gore	0.089	114.892	-3.858	-1.623	0.047	127.196	-4.812	-3.647	-5.146	0.160	68.143	-5.386	7.070	0.140	64.305	-4.249	9.525	0.025	82.899	-3.939	2.441
23	Fred Jackson	0.068	-1.025	-3.591	1.145	0.066	-37.224	-1.937	6.822	4.682	0.017	25.695	-1.099	-1.350	0.031	27.837	-1.292	-2.642	0.014	19.511	-1.089	17.770
4	Giovani Bernard	0.055	43.230	16.148	1.259	0.147	38.152	9.253	35.556	-16.010	0.066	65.527	12.289	4.869	0.064	63.035	11.488	5.958	0.166	45.786	17.287	18.099
9	James Starks	0.268	29.339	37.014	0.877	0.172	42.928	36.300	7.912	-3.937	0.273	39.727	39.614	3.977	0.273	38.922	37.329	5.351	0.392	16.828	39.452	-6.119
2	Javorius Allen	0.038	5.646	-5.842	2.073	0.172	-59.974	1.982	-28.267	32.466	0.144	77.216	-11.684	-13.627	0.170	85.893	-4.985	-19.589	0.017	62.389	-9.428	-29.314
19	Jeremy Langford	0.225	2.423	-45.759	3.630	0.372	-132.405	-46.698	-36.270	54.112	0.387	131.983	-35.217	-26.826	0.437	158.452	-44.839	-40.876	0.238	133.019	-65.019	-45.097
11	Lamar Miller	0.083	38.987	33.563	1.184	0.138	138.728	31.622	44.428	-37.361	0.086	74.120	32.283	-6.644	0.130	97.821	32.671	-23.803	0.274	126.418	53.998	-9.080
14	Latavius Murray	0.155	-15.662	26.360	3.843	0.340	-160.869	36.755	-11.361	40.172	0.374	97.797	27.188	-16.798	0.162	100.647	22.938	-17.019	0.051	89.808	4.266	-34.026
31	LeGarrette Blount	0.325	-59.848	13.471	5.095	0.469	-215.387	13.666	3.869	38.098	0.240	73.708	19.026	-12.443	0.406	99.588	16.359	-33.202	0.335	100.193	43.736	
27	LeSean McCoy	0.110	161.021	7.397	-2.920	0.570	526.528	12.125	-54.828	-30.965	0.082	80.778	8.688	9.515	0.110	76.315	10.412	11.735	0.156	64.130	15.696	20.673
3	Mike Tolbert	0.257	-26.081	9.871	1.764	0.362	-72.813	11.948	-3.325	14.497	0.138	22.564	10.821	-5.066	0.250	27.825	8.871	-8.940	0.148	10.879	4.838	5.917
13	Rashad Jennings	0.111	83.562	-20.612	-0.541	0.147	113.264	-21.280	8.675	-11.200	0.387	54.641	-33.916	17.975	0.270	52.109	-30.586	19.178	0.361	106.455	-9.699	-27.796
7	Ronnie Hillman	0.039	37.189	13.936	0.542	0.081	-212.337	26.288	26.215	21.527	0.078	62.288	18.014	-8.362	0.043	56.444	14.675	-4.034	0.036	48.931	12.831	0.671
28	T.J. Yeldon	0.046	35.380	-2.899	2.141	0.374	-37.434	-10.292	17.758	8.104	0.169	96.028	-1.596	-9.331	0.145	101.136	-2.387	-12.546	0.205	112.462	7.099	-20.811
8	Theo Riddick	0.349	29.184	20.411	0.536	0.374	81.012	19.435	-4.577	-3.031	0.404	51.038	22.606	-5.301	0.386	51.881	20.818	-5.740	0.427	19.470	28.812	10.717
26	Todd Gurley	0.255	-110.551	-45.069	10.515	0.622	-784.105	-63.994	35.697	111.582	0.083	141.182	-21.244	-14.823	0.213	179.798	-33.833	-32.413	0.231	15.321	10.296	45.149

Table 12.7 Running Back Prediction Model: Total Points Per Game

Number	Player	Game Points Model				Team Statistics Model					Logistic Probability Model				Logit Spread Model				Logit Points Model			
		R²	Const	HF	OppPAG	R²	Const	HF	OppY/R	OppY/PA	R²	Const	HF	Logistic Probability	R²	Const	HF	Logit Spread	R²	Const	HF	Logit Points
12	Adrian Peterson	0.015	6.658	-0.995	-0.104	0.051	-0.407	-1.541	3.331	-1.288	0.040	6.012	-0.900	-0.878	0.027	5.680	-0.600	-0.841	0.106	0.591	-0.214	2.321
24	Alfred Morris	0.176	-1.388	-1.471	0.114	0.235	-2.866	-1.326	1.565	-0.337	0.133	1.254	-1.223	-0.037	0.143	1.758	-1.409	-0.360	0.138	1.581	-1.081	-0.227
29	Antonio Andrews	0.059	0.407	1.364	0.034	0.076	5.167	1.439	-1.254	0.158	0.109	2.066	1.417	-0.650	0.081	2.052	1.316	-0.619	0.070	0.303	1.231	0.623
25	C.J. Spiller	0.195	-3.619	1.641	0.152	0.305	-13.145	1.675	1.601	0.964	0.211	0.669	1.520	-0.420	0.175	0.504	1.668	-0.401	0.222	-1.071	0.957	0.877
32	Carlos Hyde	0.015	0.560	1.161	0.063	0.195	-28.762	1.820	4.580	1.739	0.090	3.166	0.994	-0.529	0.090	5.925	1.018	-1.793	0.072	6.862	-1.161	-2.409
30	Charcandrick West	0.081	11.009	0.827	-0.374	0.777	27.321	0.198	-10.639	3.018	0.209	-0.733	0.724	1.922	0.076	-0.779	0.468	1.925	0.027	6.217	-0.312	-1.765
21	Chris Ivory	0.253	-4.851	3.814	0.277	0.367	-5.369	3.337	-3.834	3.336	0.233	1.563	4.374	0.130	0.238	1.003	4.491	0.581	0.263	-0.399	3.209	1.631
20	Chris Polk	0.162	2.694	-1.650	-0.043	0.172	-1.085	-1.685	0.795	-0.069	0.174	2.117	-1.747	-0.255	0.161	1.951	-1.747	-0.161	0.157	1.617	-1.745	0.071
16	Danny Woodhead	0.314	-11.665	5.887	0.554	0.304	-23.863	6.053	0.434	3.352	0.331	4.198	6.314	-2.156	0.392	7.411	5.705	-3.642	0.379	-7.631	7.937	4.382
6	Darren McFadden	0.031	1.224	-0.785	0.019	0.086	-5.773	-0.167	1.181	0.321	0.064	2.278	-0.439	-0.505	0.057	2.348	-0.396	-0.589	0.074	0.218	-0.999	1.122
15	Darren Sproles	0.098	6.506	1.643	-0.258	0.289	16.646	1.594	-2.285	-0.957	0.038	1.233	0.711	-0.210	0.031	0.970	0.824	-0.068	0.036	1.371	0.914	-0.280
1	David Johnson	0.187	-4.468	-2.696	-0.450	0.440	-15.217	-4.232	10.726	-3.273	0.096	7.124	-2.919	-0.737	0.166	9.003	-2.991	-1.924	0.127	8.300	-3.029	-1.445
22	DeAngelo Williams	0.046	7.356	2.925	-0.222	0.081	13.580	1.734	-4.008	0.916	0.064	0.688	2.782	1.081	0.037	1.673	2.617	0.572	0.049	-0.064	3.138	1.488
18	Devonta Freeman	0.061	19.574	-1.737	-0.530	0.146	11.083	-2.653	5.926	-0.955	0.021	6.668	-1.034	-0.741	0.009	6.178	-1.199	-0.189	0.030	7.910	-0.359	-1.546
17	Doug Martin	0.282	-3.343	4.998	0.139	0.347	12.074	3.626	-5.341	1.611	0.278	-0.496	5.260	0.396	0.274	-0.360	5.150	0.342	0.325	2.380	7.082	-1.979
5	Duke Johnson	0.170	0.924	-1.780	0.038	0.631	-10.349	-1.541	4.158	-0.692	0.169	1.875	-1.715	-0.090	0.167	1.631	-1.701	0.041	0.173	2.234	-1.902	-0.268
10	Frank Gore	0.129	5.101	-2.532	-0.074	0.183	9.677	-2.936	-2.723	0.681	0.064	3.747	-2.550	-0.222	0.126	3.748	-2.582	-0.218	0.218	5.456	-2.230	-1.424
23	Fred Jackson	0.251	6.909	-1.220	-0.243	0.261	-5.432	-1.480	2.919	-0.735	0.305	0.422	-1.807	0.770	0.265	0.187	-1.704	0.915	0.287	-0.597	-1.612	1.440
4	Giovani Bernard	0.129	6.997	-0.652	-0.261	0.421	3.979	-0.626	3.749	-2.660	0.025	0.367	-0.047	0.325	0.052	-0.141	-0.154	0.605	0.105	2.509	-0.262	-1.084
2	James Starks	0.127	-5.028	2.619	0.266	0.190	-15.129	1.998	3.819	0.081	0.193	4.150	1.561	-1.457	0.167	3.938	2.423	-1.683	0.392	-4.972	3.046	3.309
14	Javorius Allen	0.265	-2.289	-1.543	0.176	0.344	-8.212	-1.627	0.512	1.137	0.425	3.298	-1.963	-0.886	0.343	3.341	-1.573	-0.965	0.260	3.118	-1.946	-0.901
19	Jeremy Langford	0.194	2.430	-3.355	0.122	0.343	12.809	-4.055	-5.480	2.324	0.190	4.366	-3.387	0.320	0.188	5.635	-3.295	-0.326	0.638	16.785	-7.925	-6.376
11	Lamar Miller	0.002	2.148	0.319	0.066	0.051	7.000	0.230	2.900	-2.236	0.013	2.996	0.256	0.756	0.001	3.570	0.248	0.141	0.009	4.815	0.642	-0.815
14	Latavius Murray	0.242	-8.497	1.148	0.467	0.220	-13.958	0.295	1.768	1.304	0.192	4.028	0.272	-1.094	0.091	4.236	0.006	-1.124	0.030	3.302	-1.141	-0.461
31	LeGarrette Blount	0.386	-26.649	-1.265	1.298	0.286	-17.101	0.773	-5.934	6.633	0.109	5.963	0.685	-2.143	0.219	9.683	0.382	-5.149	0.152	9.460	4.496	-5.031
27	LeSean McCoy	0.174	10.755	-0.668	-0.357	0.005	5.979	-0.178	-0.316	-0.314	0.077	1.234	-0.448	0.952	0.136	0.638	-0.285	1.273	0.093	0.421	0.132	1.481
3	Mike Tolbert	0.178	-7.990	2.222	0.346	0.322	-5.950	-2.054	7.021	-2.932	0.068	1.168	2.023	-0.442	0.073	1.466	1.824	-0.616	0.076	-0.077	1.433	0.681
13	Rashad Jennings	0.370	18.608	-3.196	-0.667	0.300	16.177	-1.719	0.954	-2.573	0.212	0.648	-1.922	1.260	0.275	-0.188	-2.129	2.055	0.044	2.463	-0.673	-0.569
7	Ronnie Hillman	0.389	8.413	3.391	-0.324	0.353	-10.010	4.587	3.227	-0.445	0.480	-2.022	3.006	1.886	0.378	-1.487	3.495	1.401	0.378	3.479	3.405	-1.712
28	T.J. Yeldon	0.105	-4.269	-1.008	0.263	0.095	-0.182	-0.532	-1.594	1.217	0.130	2.708	-0.910	-0.682	0.113	3.060	-0.968	-0.898	0.193	4.243	-0.147	-1.786
8	Theo Riddick	0.153	-3.998	0.667	0.229	0.131	-6.536	1.032	2.308	-0.364	0.141	2.686	1.163	-0.926	0.074	2.309	0.852	-0.737	0.082	3.043	0.033	-1.110
26	Todd Gurley	0.547	-19.508	-0.168	1.069	0.581	-56.720	-0.349	1.920	7.708	0.599	9.443	1.022	-2.885	0.602	11.338	0.513	-3.873	0.406	-5.191	5.149	3.822

Table 12.8 Wide Receiver Prediction Model: Total Yards Per Game

Number	Player	Game Points Model				Team Statistics Model					Logistic Probability Model				Logit Spread Model				Logit Points Model			
		R²	Const	HF	OppPPAG	R²	Const	HF	OppY/R	OppY/PA	R²	Const	HF	Logistic Probability	R²	Const	HF	Logit Spread	R²	Const	HF	Logit Points
3	A.J. Green	0.062	115.654	−26.131	−0.738	0.176	259.233	−21.515	−59.554	12.073	0.061	96.562	−24.453	1.145	0.092	80.160	−27.076	10.995	0.063	104.572	−25.281	−4.126
9	Allen Robinson	0.052	43.371	−14.919	1.994	0.220	38.507	−0.366	−48.572	34.999	0.300	111.694	−13.068	−20.077	0.175	115.714	−15.810	−21.088	0.053	79.081	−20.518	9.361
29	Anquan Boldin	0.030	16.411	8.651	1.712	0.085	114.538	8.585	−35.426	12.400	0.036	66.853	7.632	−6.095	0.006	61.001	4.730	−2.653	0.020	40.615	11.764	7.981
16	Antonio Brown	0.085	37.940	31.063	2.772	0.105	230.401	37.389	−40.849	4.681	0.077	104.260	38.140	−5.576	0.117	138.703	26.946	−19.998	0.171	6.627	73.430	41.617
11	Brandin Cooks	0.034	22.448	0.420	2.053	0.327	63.664	14.133	52.898	−12.850	0.097	56.153	7.680	7.726	0.073	55.406	5.593	10.138	0.097	90.328	12.626	−14.565
13	Brandon Marshall	0.222	−78.562	−21.166	7.263	0.240	−205.658	−19.229	9.537	37.735	0.085	10.406	−15.943	−14.334	0.174	127.322	−16.761	−27.444	0.029	80.387	−14.146	10.284
5	Calvin Johnson	0.239	74.339	35.363	−0.951	0.384	130.656	28.468	50.000	−42.326	0.230	52.582	34.310	0.791	0.239	44.499	34.605	4.894	0.539	−23.958	64.014	39.667
27	Danny Amendola	0.003	37.791	−2.462	0.664	0.248	168.948	6.631	39.778	−41.360	0.112	36.646	−2.848	17.477	0.024	43.307	−3.830	9.642	0.237	103.258	25.675	−44.062
7	DeAndre Hopkins	0.020	58.942	1.768	1.477	0.105	212.262	−1.228	−37.956	5.366	0.123	112.466	−0.505	−12.678	0.138	116.111	0.791	−16.144	0.138	115.874	9.710	−16.972
4	Demaryius Thomas	0.052	113.790	7.264	−1.615	0.168	−112.133	17.408	56.423	−7.838	0.053	66.178	7.303	6.429	0.043	66.245	8.375	5.873	0.033	69.807	13.344	3.650
17	Doug Baldwin	0.214	199.524	18.472	−6.488	0.113	317.664	8.042	−32.804	−17.418	0.168	37.821	2.531	13.805	0.345	14.630	2.957	27.795	0.025	74.236	5.302	−8.248
31	Harry Douglas	0.126	47.612	−14.756	−0.596	0.229	17.808	−20.168	21.415	−9.840	0.200	27.048	−16.791	5.719	0.135	28.339	−14.798	3.731	0.168	19.221	−15.782	9.454
25	Jarius Wright	0.246	70.528	10.891	−2.054	0.215	55.672	12.174	10.252	−10.941	0.462	−0.875	12.652	13.313	0.393	1.150	9.297	13.669	0.199	30.866	14.129	−5.863
10	Jarvis Landry	0.176	174.480	−25.393	−3.626	0.269	293.817	−25.913	−30.303	−11.552	0.118	79.046	−21.563	8.108	0.101	78.806	−21.718	6.530	0.097	81.037	−23.864	4.660
24	Jeremy Maclin	0.027	118.826	11.773	−2.191	0.568	186.602	−25.146	−139.300	−106.386	0.011	66.744	8.527	2.924	0.085	25.338	14.362	24.573	0.041	41.319	10.486	18.296
15	Jordan Matthews	0.017	102.123	−1.619	−1.765	0.022	−6.451	−10.985	6.698	6.311	0.007	63.605	−7.069	−0.099	0.010	57.506	−5.278	3.589	0.324	128.501	0.166	−35.418
2	Julio Jones	0.225	−33.665	0.925	5.806	0.179	4.395	8.048	−29.359	31.620	0.160	128.085	−4.045	−14.517	0.234	141.071	−4.680	−27.636	0.077	95.677	−18.023	15.642
20	Kamar Aiken	0.242	37.565	34.400	0.364	0.309	13.662	29.613	31.199	−13.239	0.261	53.502	32.203	−4.277	0.244	41.779	34.365	2.418	0.307	66.287	32.918	13.116
1	Larry Fitzgerald	0.030	112.766	0.594	−1.404	0.095	129.718	5.144	−32.319	11.905	0.027	70.694	0.863	6.183	0.030	69.504	1.511	6.804	0.096	60.286	1.794	12.464
28	Malcom Floyd	0.480	96.844	−34.346	−1.647	0.486	−10.527	−33.725	22.262	−2.960	0.477	51.891	−36.194	5.277	0.492	44.213	−34.175	8.822	0.477	44.728	−36.301	10.013
32	Marques Wilson	0.068	85.087	11.040	−2.098	0.472	371.832	−2.957	14.116	−57.664	0.092	50.523	3.960	10.438	0.086	12.337	7.360	14.594	0.294	−42.878	51.725	46.726
14	Michael Crabtree	0.113	−2.947	24.383	2.385	0.153	−37.816	28.845	−19.510	31.144	0.079	58.738	18.146	−3.880	0.060	56.329	15.573	−1.861	0.553	−7.078	37.723	38.297
30	Mike Evans	0.066	113.752	−23.243	−0.615	0.110	−35.576	−23.155	24.542	4.352	0.115	114.787	−25.627	−11.738	0.065	101.426	−24.251	−1.733	0.157	71.798	−50.160	25.032
19	Pierre Garcon	0.065	43.008	−9.936	0.375	0.122	82.520	−8.274	−14.293	3.986	0.139	63.608	−15.521	−6.476	0.101	63.488	−14.268	−6.914	0.066	55.693	−7.964	−2.380
6	Randall Cobb	0.275	122.589	23.101	−3.429	0.420	238.941	28.406	−4.903	−25.871	0.319	65.230	32.855	13.184	0.375	54.587	25.510	20.374	0.161	54.587	23.085	−4.387
26	Robert Woods	0.112	−3.443	−6.086	2.107	0.073	−28.556	−11.254	17.834	0.274	0.028	43.787	−9.486	1.673	0.025	45.383	−8.636	0.270	0.121	63.925	−10.188	−13.505
12	Rueben Randle	0.277	−104.773	29.327	5.760	0.290	−120.922	18.928	25.950	6.885	0.038	90.448	10.979	−2.172	0.047	45.536	12.177	−4.901	0.034	39.083	8.671	1.456
8	T.Y. Hilton	0.491	−91.113	5.997	7.040	0.399	−179.408	7.295	−0.249	37.075	0.259	65.192	11.220	−15.847	0.289	103.414	8.525	−24.321	0.029	75.641	11.047	−5.566
18	Tavon Austin	0.029	54.559	−12.027	0.473	0.144	−37.653	−29.165	−24.794	31.144	0.027	64.979	−11.174	−0.232	0.049	77.600	−13.733	−6.067	0.269	11.230	0.036	28.592
21	Ted Ginn	0.365	−73.181	28.688	4.753	0.326	−133.268	−20.693	−1.961	1.693	0.295	64.979	31.826	−19.127	0.294	71.566	22.752	−20.883	0.206	29.374	12.902	15.522
23	Terrance Williams	0.499	6.830	−20.239	2.053	0.497	17.558	26.998	14.841	65.230	0.448	65.230	−22.944	−5.195	0.449	67.060	21.403	−7.096	0.036	63.346	−27.872	−2.937
22	Travis Benjamin	0.063	13.040	1.639	2.547	0.197	103.405	0.077	−47.114	22.760	0.132	83.764	6.416	−10.027	0.125	94.523	1.088	−14.065	0.040	48.798	13.080	8.812
33	Will Tye	0.348	150.561	−2.542	−4.500	0.416	−62.774	26.125	22.931	0.409	0.214	35.945	17.510	0.349	0.240	31.490	12.120	5.450	0.416	−4.833	14.457	37.575

Table 12.9 Wide Receiver Prediction Model: Total Points Per Game

Number	Player	Game Points Model				Team Statistics Model					Logistic Probability Model				Logit Spread Model				Logit Points Model			
		R²	Const	HF	OppPAG	R²	Const	HF	OppY/R	OppY/PA	R²	Const	HF	Logistic Probability	R²	Const	HF	Logit Spread	R²	Const	HF	Logit Points
3	A.J. Green	0.026	-2.332	0.612	0.245	0.086	-2.892	0.612	-2.859	2.594	0.047	4.842	0.137	-0.939	0.022	4.792	0.210	-0.827	0.038	1.377	0.325	1.347
9	Allen Robinson	0.231	-17.160	-0.060	0.973	0.313	-23.038	0.578	-4.109	6.560	0.314	8.847	-0.304	-2.740	0.294	10.537	-0.655	-3.846	0.072	9.052	0.337	-2.431
29	Anquan Boldin	0.041	3.205	-1.157	-0.052	0.083	2.317	-1.551	2.124	-1.292	0.037	2.074	-0.972	-0.041	0.042	1.495	-1.120	0.259	0.064	3.493	-1.811	-0.783
16	Antonio Brown	0.258	1.296	5.050	-0.015	0.286	-0.732	4.447	-2.042	1.564	0.260	1.420	4.929	-0.196	0.292	3.608	4.256	-1.125	0.294	-2.334	6.294	1.619
11	Brandin Cooks	0.178	-11.235	1.219	0.591	0.488	-37.985	2.484	7.836	1.190	0.043	3.003	4.190	-0.234	0.038	2.571	1.714	0.001	0.100	4.806	2.892	-1.636
13	Brandon Marshall	0.412	0.876	4.149	0.071	0.417	-3.314	4.070	0.482	0.564	0.411	2.757	4.190	-0.159	0.448	4.123	3.848	-1.267	0.517	5.689	5.891	-2.406
5	Calvin Johnson	0.230	-4.819	4.063	0.267	0.345	-9.918	4.280	7.838	-3.321	0.296	4.479	4.891	-1.835	0.302	5.532	4.270	-2.373	0.420	-8.409	7.779	4.706
27	Danny Amendola	0.215	-3.555	2.225	0.149	0.317	17.786	1.981	-1.636	-2.344	0.191	0.199	2.256	-0.372	0.199	0.487	2.320	-0.457	0.191	-0.370	2.041	0.329
7	DeAndre Hopkins	0.087	10.058	1.958	-0.292	0.256	33.322	1.744	-3.981	-2.010	0.040	4.016	1.612	-0.372	0.066	2.059	1.842	0.934	0.126	5.909	2.424	-1.796
4	Demaryius Thomas	0.276	7.659	-4.188	-0.165	0.351	10.212	-3.635	-4.920	2.108	0.302	2.315	-4.394	0.977	0.334	1.281	-4.584	1.546	0.421	-0.630	-2.393	2.748
17	Doug Baldwin	0.008	1.959	0.600	0.149	0.049	23.870	0.413	-6.341	1.131	0.009	5.847	1.002	-0.419	0.028	3.035	0.660	1.264	0.012	3.570	1.093	0.980
31	Harry Douglas	0.015	-1.061	0.081	0.082	0.075	6.362	0.197	-1.633	0.163	0.004	1.134	0.063	-0.136	0.008	1.314	0.046	-0.270	0.028	2.070	0.129	-0.753
25	Jarius Wright	1.000	0.000	0.000	0.000	1.000	0.000	0.000	0.000	0.000	1.000	0.000	0.000	0.000	1.000	0.000	0.000	0.000	1.000	0.000	0.000	0.000
10	Jarvis Landry	0.045	8.734	-0.530	-0.269	0.167	25.848	-0.690	0.076	-3.470	0.001	2.233	-0.250	0.017	0.036	3.871	-0.232	-1.272	0.134	5.386	0.903	-2.400
24	Jeremy Maclin	0.254	-10.528	-1.789	0.640	0.045	-4.383	-1.111	-2.344	2.625	0.055	4.696	-0.844	-0.861	0.074	6.682	-1.120	-1.918	0.006	3.298	-0.627	-0.175
15	Jordan Matthews	0.047	6.532	-0.306	-0.181	0.025	3.435	-0.889	-0.466	0.155	0.028	2.194	-0.710	0.211	0.063	1.244	-0.471	0.804	0.039	3.576	-0.746	-0.547
2	Julio Jones	0.063	-6.116	0.731	0.356	0.013	5.991	0.380	-1.335	0.333	0.225	4.842	0.979	-2.044	0.136	5.303	0.558	-2.441	0.017	1.842	-0.421	0.937
20	Kamar Aiken	0.049	6.196	-0.611	-0.159	0.025	-1.406	-0.665	0.783	0.125	0.010	2.566	-0.570	0.003	0.010	2.404	-0.569	0.099	0.136	4.918	-0.731	-1.507
1	Larry Fitzgerald	0.107	-6.020	-1.466	0.420	0.108	-17.541	-2.128	2.157	1.766	0.089	6.353	-1.563	-1.707	0.111	6.964	-1.741	-2.059	0.030	4.538	-1.760	-0.495
28	Malcom Floyd	0.205	-0.457	-3.348	0.164	0.180	3.029	-3.219	-0.633	0.394	0.217	4.267	-3.202	-0.651	0.189	4.043	-3.212	-0.512	0.182	3.814	-3.072	-0.467
32	Marquess Wilson	0.408	8.830	-1.564	-0.353	0.791	18.126	-1.833	3.598	-4.766	0.248	-0.446	-1.989	0.891	0.527	-3.129	-2.120	2.300	0.274	-2.322	0.441	1.990
14	Michael Crabtree	0.013	3.664	-0.900	-0.010	0.088	2.625	0.162	-3.301	2.054	0.070	4.532	-0.002	-0.829	0.013	3.271	-0.939	0.106	0.106	0.051	0.456	2.133
30	Mike Evans	0.045	0.331	-1.128	0.072	0.218	-5.215	0.127	4.557	-1.812	0.040	2.215	-1.018	-0.165	0.040	2.284	-0.988	-0.256	0.039	1.767	-1.218	0.211
19	Pierre Garcon	0.243	10.048	2.449	-0.408	0.436	17.287	2.858	2.221	-3.786	0.094	-0.380	2.002	0.728	0.154	-1.953	2.570	1.680	0.280	5.636	2.175	-2.530
6	Randall Cobb	0.252	19.866	-0.140	-0.783	0.481	17.315	-0.154	9.233	-7.822	0.176	-2.328	1.503	2.168	0.280	-4.120	0.322	3.656	0.098	6.364	-0.309	-2.153
26	Robert Woods	0.102	0.294	1.673	0.020	0.319	16.139	2.180	-3.565	-0.133	0.111	1.071	1.803	-0.274	0.104	0.976	1.703	-0.166	0.234	2.783	1.467	-1.510
12	Rueben Randle	0.152	12.611	-2.986	-0.363	0.183	12.996	-2.283	0.976	-1.908	0.238	2.240	-2.902	1.404	0.198	1.859	-2.764	1.696	0.216	6.213	-1.028	-2.115
8	T.Y. Hilton	0.430	-10.920	4.030	0.484	0.558	-26.882	4.218	1.223	3.292	0.275	0.847	4.342	-0.591	0.305	1.815	4.225	-1.236	0.442	-3.422	3.710	2.403
18	Tavon Austin	0.005	6.337	-0.020	-0.111	0.295	22.461	-4.172	-9.110	3.099	0.005	3.296	-0.073	0.311	0.004	4.724	-0.401	-0.340	0.102	-1.038	0.794	2.696
21	Ted Ginn	0.062	3.511	-2.396	0.096	0.098	-6.232	-2.596	0.683	1.312	0.063	6.410	-2.248	-0.524	0.091	7.429	-2.389	-1.384	0.058	5.851	-2.623	0.112
23	Terrance Williams	0.307	-1.771	-1.937	0.169	0.307	-5.768	-1.700	0.378	0.918	0.288	3.121	-2.099	-0.492	0.316	3.561	-1.725	-0.920	0.262	3.121	-2.564	-0.412
22	Travis Benjamin	0.309	-9.035	1.925	0.477	0.203	-2.030	2.775	-1.645	1.426	0.501	4.207	2.820	-1.876	0.397	5.430	2.018	-2.243	0.453	-5.569	5.003	3.319
33	Will Tye	0.208	20.462	-3.408	-0.747	0.059	12.206	-0.838	-0.466	-1.231	0.219	0.187	-2.114	1.506	0.114	0.179	-1.631	1.512	0.348	8.695	0.621	-6.580

Table 12.10 Quarterback Prediction Model: Total Yards Per Game

		Game Points Model				Team Statistics Model					Logistic Probability Model				Logit Spread Model				Logit Points Model			
Number	Player	R^2	Const	HF	OppPAG	R^2	Const	HF	OppY/R	Opp Y/PA	R^2	Const	HF	Logistic Probability	R^2	Const	HF	Logit Spread	R^2	Const	HF	Logit Points
13	Aaron Rodgers	0.094	248.078	41.387	-0.230	0.148	187.043	37.029	58.092	-26.880	0.151	204.596	53.210	16.993	0.139	206.253	43.199	20.080	0.101	257.208	40.272	-8.065
11	Alex Smith	0.311	467.067	-25.504	-9.202	0.439	342.076	-12.385	132.165	-95.024	0.159	243.414	-35.754	14.773	0.327	175.248	-32.680	50.670	0.186	208.094	-28.592	37.792
21	Andy Dalton	0.118	400.868	-59.177	-4.882	0.209	579.215	-54.464	22.878	-56.390	0.104	269.767	-48.260	10.591	0.109	258.109	-49.215	15.827	0.116	329.929	-58.461	-25.324
25	Ben Roethlisberger	0.030	289.911	15.483	1.917	0.332	605.004	29.988	-126.854	34.693	0.025	319.159	25.052	2.943	0.028	310.012	29.048	6.653	0.474	126.608	99.813	91.412
12	Blake Bortles	0.039	362.665	-25.166	-2.290	0.050	217.976	-19.553	-12.348	20.143	0.084	322.784	-21.253	-14.069	0.104	336.569	-22.945	-24.072	0.033	302.282	-25.637	4.687
26	Brian Hoyer	0.244	92.794	26.397	6.899	0.115	233.385	-19.089	-39.022	25.823	0.307	286.200	-12.907	-30.912	0.361	299.569	-18.196	-36.522	0.097	275.873	-11.749	-19.830
15	Cam Newton	0.019	214.148	9.303	2.644	0.369	-83.700	-55.346	126.918	-19.582	0.133	257.767	-18.330	34.040	0.078	253.257	0.317	25.264	0.062	246.931	-5.055	25.343
17	Carson Palmer	0.136	454.468	9.179	-6.585	0.143	535.155	23.877	-84.855	16.510	0.032	282.018	12.232	12.689	0.091	261.937	13.512		0.059	269.468	14.031	20.045
29	Colin Kaepernick	0.217	-70.943	8.709	14.075	0.556	-1187.670	85.074	-234.390	343.234	0.324	359.554	-11.778	-56.330	0.189	361.207	-8.229	-58.553	0.209	65.073	53.620	94.146
5	Derek Carr	0.307	-119.070	93.814	15.560	0.582	-406.196	130.362	-126.445	170.806	0.395	312.993	76.028	-47.521	0.346	355.943	82.100	-71.689	0.042	217.823	33.939	20.138
18	Drew Brees	0.347	30.381	68.254	10.470	0.330	-172.540	63.171	19.282	54.783	0.228	293.394	67.343	-7.599	0.216	279.185	70.415	2.228	0.290	240.349	36.702	35.999
7	Eli Manning	0.121	68.039	69.777	7.641	0.290	-164.721	68.422	-20.888	71.487	0.084	254.546	36.124	8.130	0.078	265.936	46.015	-4.871	0.205	333.247	60.690	-54.547
2	Jameis Winston	0.280	410.967	-15.247	-5.904	0.247	407.133	-8.461	30.298	-38.563	0.253	243.653	-15.563	19.890	0.364	225.513	-20.531	40.944	0.137	293.372	-1.216	-20.620
19	Jay Cutler	0.295	90.603	-41.210	9.140	0.131	130.504	-31.406	26.240	5.824	0.336	349.387	-25.416	-33.507	0.317	368.492	-37.234	-43.891	0.119	241.612	-19.709	22.391
27	Joe Flacco	0.167	107.786	34.385	6.918	0.227	-64.810	11.431	94.491	-6.457	0.169	313.771	16.013	-23.618	0.145	309.932	31.367	-25.483	0.455	385.006	34.859	-102.964
28	Johnny Manziel	0.107	-147.044	-136.157	18.381	0.286	-486.974	-151.075	81.652	155.821	0.089	259.250	-69.065	-23.017	0.062	253.910	-68.814	-21.840	0.357	-19.354	31.588	72.029
1	Kirk Cousins	0.573	-47.737	-9.160	13.221	0.308	-105.618	8.191	48.990	22.957	0.143	281.322	10.280	-18.813	0.154	298.267	6.192	-29.920	0.108	281.947	36.945	-19.802
23	Marcus Mariota	0.378	-173.376	42.720	16.721	0.312	-298.244	67.655	-24.586	89.362	0.266	289.355	46.334	-43.958	0.425	338.093	39.572	-79.245	0.032	273.415	29.441	-19.159
32	Matt Cassel	0.702	-321.296	36.558	20.758	0.706	-1386.319	145.349	47.847	192.039	0.332	247.048	1.232	-43.108	0.393	274.314	11.512	-65.362	0.130	118.961	36.933	51.267
30	Matt Hasselbeck	0.740	-88.045	0.200	13.558	0.663	159.807	14.648	-92.208	61.445	0.633	290.958	-1.599	-56.818	0.558	303.540	-3.082	-62.151	0.087	207.716	29.151	6.046
16	Matt Ryan	0.332	59.051	26.463	8.746	0.381	-9.084	41.617	-36.657	60.689	0.201	301.400	18.281	-20.414	0.244	315.740	16.437	-34.705	0.070	258.962	-0.237	19.459
14	Matthew Stafford	0.300	113.246	44.941	6.251	0.255	62.578	54.448	57.624	-9.246	0.413	316.592	61.942	-35.757	0.441	339.490	49.819	-47.445	0.458	112.201	100.582	67.956
24	Nick Foles	0.498	381.266	49.999	-10.240	0.526	631.554	33.395		-47.686	0.231	259.419	-50.992	16.455	0.286	273.941	38.573	25.346	0.096	179.424	26.159	-3.356
31	Peyton Manning	0.154	289.992	-42.281	-0.284	0.667	-707.740	-29.734	270.081	-23.176	0.071	350.578	-11.254	-19.812	0.160	376.467	-44.474	6.242	0.295	336.733	-49.515	-36.931
4	Philip Rivers	0.198	114.542	21.858	9.378	0.361	-340.342	-20.775	39.361	74.347	0.223	265.493	26.768	5.236	0.099	258.463	-16.343	-31.679	0.085	248.302	2.657	36.574
3	Russell Wilson	0.228	324.038	32.589	-2.330	0.338	515.986	25.911	-57.546	-0.184	0.015	278.967	-8.509	-14.324	0.246	284.131	27.095	9.484	0.192	272.137	28.893	1.339
6	Ryan Fitzpatrick	0.008	190.692	-5.626	3.018	0.100	-80.527	-15.048	-11.908	56.953	0.221	275.014	11.904	-3.027	0.014	273.093	-6.016	-17.839	0.009	241.167	-10.728	16.296
10	Ryan Tannehill	0.248	407.727	-68.588	-5.017	0.322	630.409	-70.405	-43.436	-23.416	0.143	232.108	55.106	11.904	0.212	258.230	-63.529	10.819	0.224	309.050	-55.221	-16.973
20	Sam Bradford	0.274	-10.259	11.067	11.067	0.274	-248.944	23.958	88.466	17.368	0.397	155.829	-44.841	46.902	0.177	180.223	49.333	-18.873	0.289	300.126	62.612	-39.875
9	Teddy Bridgewater	0.141	363.242	-53.820	-5.054	0.141	321.466	-58.349	49.423	40.147	0.014	328.008	1.983	-6.876	0.241	323.304	-59.663	40.895	0.103	251.812	-47.806	-2.395
8	Tom Brady	0.059	217.528	-0.488	4.422	0.018	219.615	4.024	14.643	5.734					0.004		4.341	-3.702	0.014	300.864	-1.953	13.734
22	Tyrod Taylor	0.011	253.452	-10.491	0.503	0.557	-60.168	-8.086	126.927	-29.233	0.110	241.285	-9.434	17.913	0.085	239.401	-6.388	17.383	0.010	263.155	-10.181	1.165

Table 12.11 Quarterback Prediction Model: Total Points Per Game

Number	Player	Game Points Model				Team Statistics Model					Logistic Probability Model				Logit Spread Model				Logit Points Model			
		R²	Const	HF	OppPAG	R²	Const	HF	OppY/R	OppY/PA	R²	Const	HF	Logistic Probability	R²	Const	HF	Logit Spread	R²	Const	HF	Logit Points
13	Aaron Rodgers	0.075	25.017	1.609	−0.589	0.317	16.154	1.085	13.554	−8.857	0.104	6.651	3.355	2.367	0.170	4.390	2.081	4.158	0.028	14.227	1.533	−1.264
11	Alex Smith	0.128	−8.235	−0.497	0.744	0.002	5.723	0.337	−0.540	0.616	0.072	10.834	0.262	−1.693	0.208	17.189	−0.013	−5.048	0.010	9.624	0.195	−1.246
21	Andy Dalton	0.306	−13.255	−0.972	1.206	0.250	−25.406	−3.198	1.935	4.718	0.330	20.738	−3.597	−3.632	0.261	22.486	−3.405	−4.173	0.074	13.234	−3.411	1.030
25	Ben Roethlisberger	0.716	−10.716	10.194	0.724	0.787	2.759	12.298	−7.909	4.913	0.669	6.539	11.934	−1.469	0.730	10.649	10.112	−3.134	0.772	−8.727	17.517	5.369
12	Blake Bortles	0.145	−3.697	−2.663	0.828	0.146	1.567	−1.801	−4.435	4.608	0.181	18.431	−2.869	−2.340	0.224	20.637	−3.153	−3.930	0.105	19.599	−1.968	−2.867
26	Brian Hoyer	0.647	−12.725	−5.955	1.139	0.520	−13.064	−5.379	−4.266	6.400	0.706	18.793	−3.802	−4.750	0.723	20.299	−4.628	−5.209	0.138	13.122	−4.823	0.060
15	Cam Newton	0.072	5.581	3.993	0.385	0.440	−25.307	−4.327	15.881	−3.138	0.104	13.575	1.590	2.635	0.069	13.602	3.101	1.850	0.045	15.084	3.325	0.251
17	Carson Palmer	0.006	11.008	−0.156	0.139	0.202	−11.455	−1.474	10.296	−2.382	0.000	14.322	−0.245	−0.047	0.017	12.498	−0.255	1.123	0.001	14.068	−0.248	0.115
29	Colin Kaepernick	0.559	−18.562	−2.183	1.167	0.680	−69.133	1.572	−15.575	20.146	0.406	13.617	−4.106	−2.969	0.256	12.575	−4.025	−2.525	0.380	−3.653	0.055	5.885
5	Derek Carr	0.026	7.415	−0.573	0.266	0.205	1.990	2.182	−9.571	7.286	0.128	16.668	0.574	−2.218	0.064	17.239	0.114	−2.379	0.171	5.435	1.505	5.228
18	Drew Brees	0.455	−36.560	9.665	1.854	0.418	−72.264	9.999	7.397	7.214	0.247	9.537	9.622	−1.025	0.235	8.484	9.953	−0.383	0.242	9.836	11.474	−1.586
7	Eli Manning	0.095	−12.424	0.428	1.101	0.124	−24.992	−1.003	1.779	4.531	0.231	11.145	−7.708	5.060	0.082	12.142	−5.627	3.551	0.128	22.943	−1.578	−5.707
2	Jameis Winston	0.128	10.395	−4.393	0.102	0.192	15.215	−2.491	5.262	−3.612	0.129	12.336	−4.163	0.416	0.133	11.819	−4.264	0.988	0.126	13.310	−3.916	−0.377
19	Jay Cutler	0.167	10.728	−4.150	0.013	0.372	1.155	−2.997	8.275	−3.797	0.216	13.592	−3.730	−1.308	0.170	11.884	−4.156	−0.454	0.209	6.198	−2.253	2.598
27	Joe Flacco	0.760	−16.085	3.881	1.077	0.583	−27.808	3.652	0.996	4.675	0.709	15.369	1.365	−3.423	0.574	14.570	3.667	−3.575	0.335	9.434	5.770	−1.261
28	Johnny Manziel	0.237	−14.281	−3.714	1.002	0.125	4.971	0.147	−3.394	2.002	0.431	9.711	−0.814	−2.184	0.211	9.143	−0.746	−1.863	0.842	−6.300	5.073	5.140
1	Kirk Cousins	0.299	−7.328	6.102	0.646	0.274	−7.409	7.380	3.943	−0.230	0.247	6.112	8.484	0.469	0.245	6.520	8.215	0.273	0.248	8.682	8.335	−0.918
23	Marcus Mariota	0.564	−39.194	0.020	2.040	0.438	−37.687	4.012	−9.134	11.986	0.278	16.414	0.061	−4.492	0.425	21.122	−0.670	−7.864	0.016	10.360	−2.197	1.153
32	Matt Cassel	0.753	−34.813	9.996	1.476	0.673	−93.043	15.370	5.236	10.109	0.283	4.517	6.701	−2.260	0.423	7.181	8.001	−4.378	0.368	−7.738	5.000	6.714
30	Matt Hasselbeck	0.210	20.418	1.849	−0.577	0.071	14.091	0.855	1.701	−1.956	0.077	5.413	1.430	1.577	0.004	7.370	0.552	0.092	0.061	5.044	−0.345	1.848
16	Matt Ryan	0.300	−12.865	3.823	0.766	0.277	−18.807	4.680	−2.096	4.690	0.254	8.634	3.251	−2.090	0.384	10.650	3.195	−4.134	0.060	5.937	1.955	0.658
14	Matthew Stafford	0.454	−16.559	2.377	1.264	0.523	−44.849	5.052	11.664	0.750	0.484	21.317	5.272	−5.588	0.618	26.154	3.374	−8.053	0.140	1.577	6.712	4.423
24	Nick Foles	0.069	7.809	−2.814	−0.089	0.405	36.006	−9.241	−14.078	4.730	0.116	2.515	−2.051	1.358	0.098	3.084	−2.497	1.271	0.088	9.108	−4.114	−1.575
31	Peyton Manning	0.635	34.807	−10.375	−1.055	0.422	17.605	−7.853	4.701	−4.072	0.518	5.152	−9.369	3.078	0.485	4.307	−9.281	3.569	0.363	6.908	−7.199	1.857
4	Philip Rivers	0.717	−23.052	−4.093	1.665	0.632	−35.788	−4.249	−5.853	10.942	0.673	23.181	−2.660	−5.742	0.633	27.254	−3.652	−7.496	0.012	11.029	−1.349	0.558
3	Russell Wilson	0.002	10.079	−0.224	0.130	0.133	50.068	−1.117	−14.446	3.371	0.024	14.968	0.433	−1.257	0.000	12.613	−0.023	0.147	0.013	10.075	0.417	1.734
6	Ryan Fitzpatrick	0.421	20.482	7.817	−0.502	0.398	3.154	6.545	−1.273	1.553	0.385	7.924	7.193	0.556	0.383	9.147	6.695	−0.470	0.382	9.044	7.100	−0.365
10	Ryan Tannehill	0.113	16.663	3.387	−0.349	0.222	43.823	3.038	−3.327	−3.137	0.107	9.462	3.741	−1.217	0.180	12.762	3.800	−3.542	0.243	14.036	5.878	−4.429
20	Sam Bradford	0.128	−5.754	−2.171	0.642	0.332	−23.756	−2.065	7.015	0.482	0.001	8.233	−0.081	−0.129	0.000	8.199	−0.048	−0.119	0.095	11.474	0.265	−1.886
9	Teddy Bridgewater	0.203	−17.914	4.765	1.011	0.107	−9.102	2.183	6.258	−1.622	0.053	5.025	3.432	0.060	0.072	7.700	3.901	−1.543	0.237	−3.755	4.976	5.590
8	Tom Brady	0.034	23.054	0.857	−0.309	0.041	24.419	0.260	−2.938	0.550	0.076	13.863	1.094	1.546	0.043	13.942	0.622	1.586	0.004	15.366	0.264	0.494
22	Tyrod Taylor	0.207	31.815	3.734	−0.949	0.154	40.477	3.701	1.098	−5.139	0.066	8.373	2.965	1.454	0.088	7.150	3.366	2.141	0.356	21.622	3.742	−7.761

for different points-scoring systems. Additionally, these techniques can be used to compute the probability that a player will score at least a specified number of points or score points within a specified interval using regression results and our probability models.

It is important to point out here that the selection of the best fantasy sports team will be different based on the scoring system selected by the fantasy sports league or commissioner. However, the approaches provided in this chapter provide the necessary insight and foundation to build successful fantasy sports models.

CHAPTER 13

Advanced Modeling Techniques

13.1 INTRODUCTION

In this chapter we show how to apply advanced mathematical techniques to sports modeling problems. The techniques discussed in this chapter include principal component analysis (PCA), neural networks (NNETs), and adaptive regression analysis.

In particular, these techniques can be used by sports professionals to:

1. Compute the most appropriate set of explanatory factors for a prediction model, including regression models and probability models;
2. Determine the most optimal mix of players for a game or team, e.g., based on player rankings and/or based on the opponent's roster;
3. Adapt model parameter values in real time based on a changing mix of players available for a game—e.g., revise the model parameters based on whether a player will or will not be available for a game (due to an injury or coming off an injury).

Content in this chapter provides all sports professionals and analysts, including owners, coaches, general managers, and the amateur and professional fantasy sports competitor, with tools to fine tune their predictive models based on real-time information, player availability, and opponent rosters, as well as to determine the optimal complementary relationships across team members.

These techniques can further be used as the basis for:

- Salary Arbitration
- National Team Selection
- Hall of Fame Evaluation
- Team Trades

13.2 PRINCIPAL COMPONENT ANALYSIS

This section provides an application of PCA to sports modeling data. Readers interested in a more mathematical discussion of PCA are referred to articles on PCA, eigenvalue–eigenvector decomposition, and/or singular value decomposition.

Optimal Sports Math, Statistics, and Fantasy.
DOI: http://dx.doi.org/10.1016/B978-0-12-805163-4.00016-5

PCA is a statistical process that is used to reduce the number of data variables from a correlated set explanatory factors into a smaller subset of data variables that are uncorrelated and independent. The set of uncorrelated principal components are determined in a way such that the first principal component factor explains the greatest variability of the correlated data, the second principal component factor explains the second greatest amount of variability of the correlated data, and so on, with the last principal factor explaining the least amount of variability of the correlated data. The number of principal components will always be less than or equal to the number of original factors (variables).

PCA is an important statistical tools used in data analysis and in mathematical models. It has recently become a staple for financial modeling and algorithmic trading, and is just starting to make its way into professional sports modeling problems and sports statistics. Although many of the professional sports teams using PCA techniques claim they are not appropriate for sports and that other teams should not waste time or resources experimenting with these approaches, these claims are enough to justify their usage; otherwise, why would a team try to help a competitor.

The mathematics behind PCA is based on either eigenvalue—eigenvector decomposition (if the data set is a square matrix) or based on singular value decomposition of any data matrix (e.g., singular value decomposition can be applied to a square and rectangular matrix). Knowledge of factor analysis and matrix theory is also beneficial for PCA.

The goal of PCA is to reduce the underlying correlated data set into a smaller set of uncorrelated factors. In sports modeling problems, many times the explanatory factors are correlated, and if used in a regression analysis, will lead to erroneous conclusions and incorrect predictions. Recall from Chapter 2, Regression Models, that one of the underlying requirements for a proper analysis is to have an uncorrelated (e.g., independent) explanatory factor.

Example 13.1 PCA With Baseball

Baseball is a sport with an almost unlimited amount of data, and the number of data points seems to be increasing each season. And in many cases, these data points are highly correlated with each other. For example, for offensive hitting statistics we have: (1) batting average (AVG), (2) on-base percentage (OBP), (3) slugging percentage (SLG), and (4) a statistic that is the sum of on-base percentage plus slugging percentage appropriately titled on-base percentage plus slugging percentage (OPS), etc. But are all of these statistics necessary? What incremental value can each of these statistics provide considering that they are highly correlated?

Table 13.1A Baseball Statistics

Team	Team Record			Hitting Statistics				PCA Factors			
	Win	Loss	WinPct	AVG	OBP	SLG	OPS	S_1^*	S_2	S_3^*	S_4
Arizona Diamondbacks	79	83	.488	0.264	.324	0.414	0.738	0.021	0.007	0.003	0.000
Atlanta Braves	67	95	.414	0.251	.314	0.359	0.674	−0.064	0.016	−0.004	−0.001
Baltimore Orioles	81	81	.500	0.250	.307	0.421	0.728	0.012	−0.016	0.004	0.000
Boston Red Sox	78	84	.481	0.265	.325	0.415	0.740	0.024	0.008	0.003	0.000
Chicago Cubs	97	65	.599	0.244	.321	0.398	0.719	−0.006	0.000	−0.012	0.000
Chicago White Sox	76	86	.469	0.250	.306	0.380	0.686	−0.044	0.000	0.003	0.000
Cincinnati Reds	64	98	.395	0.248	.312	0.394	0.706	−0.020	−0.003	−0.002	0.000
Cleveland Indians	81	80	.503	0.256	.325	0.401	0.725	0.003	0.008	−0.005	0.001
Colorado Rockies	68	94	.420	0.265	.315	0.432	0.748	0.038	−0.007	0.011	−0.001
Detroit Tigers	74	87	.460	0.270	.328	0.420	0.748	0.034	0.011	0.005	0.000
Houston Astros	86	76	.531	0.250	.315	0.437	0.752	0.041	−0.018	−0.001	0.000
Kansas City Royals	95	67	.586	0.269	.322	0.412	0.734	0.018	0.009	0.008	0.000
Los Angeles Angels	85	77	.525	0.246	.307	0.396	0.702	−0.023	−0.008	0.000	0.001
Los Angeles Dodgers	92	70	.568	0.250	.326	0.413	0.739	0.020	0.001	−0.010	0.000
Miami Marlins	71	91	.438	0.260	.310	0.384	0.694	−0.034	0.007	0.008	0.000
Milwaukee Brewers	68	94	.420	0.251	.307	0.393	0.700	−0.026	−0.004	0.003	0.000
Minnesota Twins	83	79	.512	0.247	.305	0.399	0.704	−0.020	−0.010	0.002	0.000
New York Mets	90	72	.556	0.244	.312	0.400	0.712	−0.012	−0.007	−0.005	0.000
New York Yankees	87	75	.537	0.251	.323	0.421	0.744	0.028	−0.004	−0.006	0.000
Oakland Athletics	68	94	.420	0.251	.312	0.395	0.707	−0.018	−0.001	0.000	0.000
Philadelphia Phillies	63	99	.389	0.249	.303	0.382	0.684	−0.046	−0.003	0.004	0.001
Pittsburgh Pirates	98	64	.605	0.260	.323	0.396	0.719	−0.004	0.011	0.000	0.000
San Diego Padres	74	88	.457	0.243	.300	0.385	0.685	−0.045	−0.010	0.002	0.000
San Francisco Giants	84	78	.519	0.267	.326	0.406	0.732	0.013	0.013	0.004	0.000
Seattle Mariners	76	86	.469	0.249	.311	0.411	0.722	0.002	−0.010	0.000	0.000
St. Louis Cardinals	100	62	.617	0.253	.321	0.394	0.716	−0.009	0.007	−0.005	−0.001
Tampa Bay Rays	80	82	.494	0.252	.314	0.406	0.720	−0.001	−0.004	0.000	0.000
Texas Rangers	88	74	.543	0.257	.325	0.413	0.739	0.021	0.004	−0.004	−0.001
Toronto Blue Jays	93	69	.574	0.269	.340	0.457	0.797	0.096	0.003	−0.002	0.000
Washington Nationals	83	79	.512	0.251	.321	0.403	0.724	0.002	0.002	−0.006	0.000

The "*" in S_1^* and S_3^* indicate that the statistical factors S_1 and S_3 are statistically significant predictors of team winning percentage. The statistical factors S_2 and S_4 were not found to be statistically significant predictors of team winning percentage.

To determine which of these "statistics" are the most important to assess a player's offensive ability and to be used as a predictor of team's winning success, we turn to PCA for insight.

Table 13.1A shows these four statistics for each team for the 2015 season. The correlation of these statistics is quite high and is shown in Table 13.1B.

To determine a subset of data that is uncorrelated we turn to PCA. We can perform PCA analysis using a statistical package such as MATLAB's PCA function as follows:

$$[Coeff, Score, Latent, Tsquared, Explained, Mu] = pca(X)$$

where

X = original correlated data matrix
$Coeff$ = principal component weightings matrix
$Score$ = principal component factors
$Latent$ = principal component variances
$Tsquared$ = Hotelling's T-squared statistic for each observation in X
$Explained$ = percentage of total variance explained
Mu = mean of each original factor in the X data matrix

After performing our PCA analysis on data set X we have an uncorrelated Score matrix consisting of four PCA factors $S = [S_1, S_2, S_3, S_4]$ one for each of the columns of X.

The correlation matrix of Score is shown in Table 13.1C and as expected is the identity matrix representing an independent and uncorrelated data set. The four PCA factors from our analysis are shown in Table 13.1A. The matrix showing the correlation between PCA factors (Score) and the X data matrix is shown in Table 13.1D.

The percentage of total variance of data set X explained by the principal factors is shown in Table 13.1D. In this example, the first principal component explains 91.2% of the variance in X, and the second principal component factor explains 6.4% of the variance in X. Thus, combined, we have two factors that explain 97.6% of the variance in X (Table 13.1E).

Table 13.1B Correlation of Baseball Statistics

	AVG	OBP	SLG	OPS
AVG	1.00	0.67	0.47	0.60
OBP	0.67	1.00	0.58	0.80
SLG	0.47	0.58	1.00	0.96
OPS	0.60	0.80	0.96	1.00

Table 13.1C Correlation of Baseball Principal Components

	S_1	S_2	S_3	S_4
S_1	1	0	0	0
S_2	0	1	0	0
S_3	0	0	1	0
S_4	0	0	0	1

Table 13.1D Correlation of PCA Factors and X Data

S/X	AVG	OBP	SLP	OPS
S_1	0.60	0.76	0.97	1.00
S_2	0.61	0.58	-0.24	0.03
S_3	0.51	-0.28	0.07	-0.04
S_4	0.00	0.02	0.01	-0.01

Table 13.1E PCA Variance Explained

PCA	Pct	Cumulative
S_1	91.2%	91.2%
S_2	6.4%	97.6%
S_3	2.3%	100.0%
S_4	0.0%	100.0%

Table 13.1F Winning Percentage as a Function of PCA Factor Regression

Factor	Value	SE	t-Stat
Const	0.5000	0.0096	51.8742
S_1	**0.8513**	**0.3013**	**2.8256**
S_2	0.8924	1.1331	0.7875
S_3	**-5.8059**	**1.8787**	**-3.0904**
S_4	-2.3920	35.5034	-0.0674
R^2	0.4208		
MSE	0.0028		
F-Stat	4.5398		

The bolded values in Table 13.1F indicate that the factor is statistically significant.

The original data matrix X can be computed in terms of the PCA statistics as follows:

$$X = Score * Coeff' + repmat(Mu, N, 1)$$

where

N = number of data observations
$Repmat(Mu,N,1)$ = matrix where each column is the corresponding column mean of X. This notation simply entails adding the average of each column to each data point.

The next question that naturally arises is what do these principal factors mean? And can they be used to predict the winning team.

To answer these questions, we run a regression of team winning percentage "Y" as a function of the four principal component factors "S." This regression is as follows:

$$WinPct = \alpha_0 + \alpha_1 S_1 + \alpha_2 S_2 + \alpha_3 S_3 + \alpha_4 S_4 + \varepsilon$$

The results of this regression are shown in Table 13.1F and the percentage of *WinPct* explained by each statistical factor is shown in Table 13.1G.

Table 13.1G Variance Explained by Each PCA Factor	
Factor	R^2
S_1	0.1850
S_2	0.0144
S_3	0.2213
S_4	0.0001
Total	0.4208

Notice that this regression has goodness of fit of $R^2 = 42.8\%$, which is quite high for sports models. Furthermore, we see that only principal component factors S_1 and S_3 are significant explanatory factors of team winning percentage. Factors S_2 and S_4 have insignificant t-Stats. This tells us that we only need two of the four PCA factors to model team winning percentage. Furthermore, the percentage of total R^2 explained by factors S_1 and S_3 are 0.1850 and 0.2213 respectively. Factors S_2 and S_4 explain an insignificant amount of total R^2 of 0.0144 and 0.0001 respectively.

The next step in our analysis is to infer a meaning of these PCA statistical factors via a regression of the PCA factor on the underlying data set X. Since we have two significant PCA factors, we need to run two regression models of the form:

$$S = b_0 + b_1 x_1 + b_2 x_2 + b_3 x_3 + b_4 x_4 + \varepsilon$$

The results of these regressions provide us with the significant baseball hitting statistics. Having a relationship between the statistical factors and the real data set allows us to calculate the value of the statistical factors directly. These are:

$$S_1 = -0.8992 + 0.1508 \cdot AVG + 0.2151 \cdot OBP + 0.5653 \cdot SLG + 0.7820 \cdot OPS + \varepsilon$$

$$R^2 = 0.9999$$

$$S_3 = -0.0033 + 0.8020 \cdot AVG - 0.4864 \cdot OBP + 0.2702 \cdot SLG - 0.2165 \cdot OPS + \varepsilon$$

$$R^2 = 0.9999$$

In both cases, as expected, we have a very high goodness of fit between the PCA factor and the original data variables. In fact, using all four PCA factors would result in a goodness of fit of $R^2 = 1$. The significant factors S_1 and S_3 are highlighted with an asterisk in Table 13.1A. Readers can use this data table to verify our results.

In many cases, practitioners give these PCA factors a qualitative name based on their data composition. We leave these naming conventions to our readers.

13.3 NEURAL NETWORK

NNET, also known as an artificial neural network, is a process which consists of mapping an input vector to a set of nodes and then to an output value. At each node, a weighting is applied to the data point similar to how we

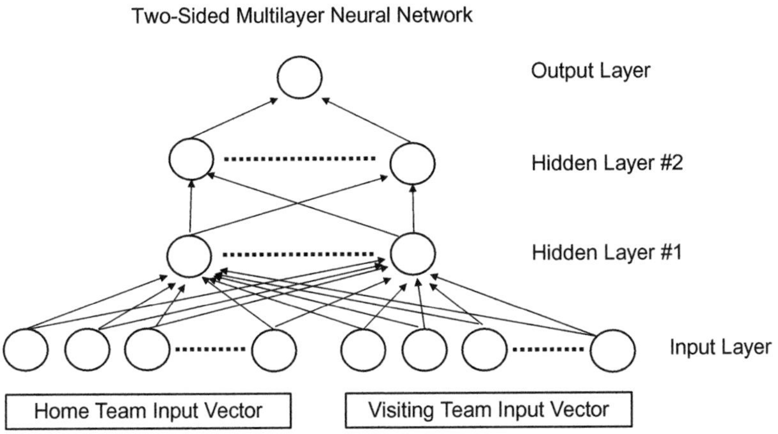

Two-Sided Multilayer Neural Network

Figure 13.1 Two-Sided Multilayer Neural Network.

apply the parameter value in our logit regression model. These new data values are then mapped to another set of nodes and the process is repeated until we arrive at the last node of the network and the output value.

NNETs have been used in industry as the basis for handwriting recognition and data mining. In the case of handwriting recognition, the input and output data values consist of binary data points, i.e., 0 or 1. NNETs also serve as the computational engine behind machine-learning algorithms and data reduction problems. The NNET determines the optimal set of weightings to apply at each node based on a method known as "learning" or "training."

In this section we show how NNETs can be applied to sports modeling problems to determine the optimal mix of players for a team and appropriate player rankings. These modeling techniques can also be expanded to calculate the expected winner of a game as well as the home team point spread. Readers interested in learning more about NNET models are encouraged to perform additional research regarding the different learning approaches and different applications of these networks.

In our sports modeling approach, we follow the techniques developed in handwriting recognition using binary data. Two different sports modeling NNET architectures for player rankings are shown in Fig. 13.1 (two-sided multilayer neural network) and Fig. 13.2 (one-sided multilayer neural network).

The two-sided multilayer neural network consists of an input vector that includes a variable for each player for the home team and another variable for each player for the away team. That is, the length of the input

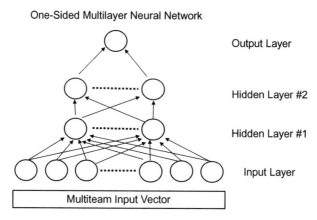

One-Sided Multilayer Neural Network

Figure 13.2 One-Sided Multilayer Neural Network.

vector is $2 \times$ the number of players. If the player is on the home team the value of the input variable is $+1$ and if they are not on the team the value of their input variable is 0 (similarly for the away team). For the one-sided multilayer neural network, the input vector consists of a variable for each player. The value of this input variable is $+1$ if the player is on the home team, -1 if the player is on the away team, and 0 if the player is not on any of the teams. Thus, the weightings at each input variable node represent the player's strength or rating.

Example 13.2 NNET Player Selection

In this example, we show how a NNET model can be used to assist teams and coaches in evaluating and ranking players. This provides coaches with an alternative means for player evaluation.

The data used for this analysis is based on a soccer tryout where players were evaluated in part on how well they performed in small-sided 6-versus-6 games. The data set included 30 players and 66 games in total. Ultimately, players were assigned to one of two teams. The more highly ranked players were assigned to the A team and the next level of ranked players were assigned to the B team.

While this example examines a soccer player's performance and rankings, the NNET techniques can be applied to almost any team sport including basketball, hockey, cricket, rugby, etc.

The NNET model used in this example is based on the one-sided multilayer network shown in Fig. 13.2. Our input vector consisted of 30 variables (one variable per player). Players on a team were arbitrarily assigned as either the home team or the away team. The output score was always taken from the perspective of the home team. The six players assigned to the home team had an input value of $+1$. The six players assigned to the away team had an input value of -1. Players who were not a part of the game had an input value of 0. The output value was defined as the home team spread, i.e., the home team score minus the away team score. If the home team won by a score of 3−1 then the output value was +2. If the home team lost by a score of 4−1 the output value was −3. If the game ended in a tie the output value was 0.

For example, if players 1−6 were assigned to the home team and players 7−12 were assigned to the away team, the input vector for this game would be +1 for the first six values, −1 for the next six values, and 0 for the remaining 18 values. If the home team won the game 2−0, the output value corresponding to this vector would be +2.

It is important to note that there are many different ways of specifying the network structure, including number of hidden layers and number of nodes at each layer, and also using different values for the input and output values. For example, based on the data, it may be appropriate to scale or transform the input and output values. The more common transformation techniques include transforming the value into a z-score and scaling the value based on the range of the data points. Additionally, for some sports, the output value could also be defined as home team time of possession, etc.

We determined player strength ratings, i.e., rankings, via our NNET model. These results were then compared to the coaches' selection of players using in-sample data. Our data observations were not large enough to perform a full in-sample and out-sample analysis of the results, but the approach described can be applied to a larger data sample with improved accuracy.

For each of the 66 small-sided games, we estimated the winning team based on our ranking of the players. We also estimated the winning team based on the coach's selection of players. If in a game, the home team had more A-Team players than the away team, then the coaches model would predict the home team to win. If the away team had more A-Team players than the home team did, then the coaches model would predict the away team to win. If the home and away team had the same number of A players, then the coaches model would predict a tie. In actuality, the coaches model could further fine tune these predictions based on the coaches ranking of each individual player. However, our data set only included the end result where the players was either assigned to the A team or the B team.

In the case of a predicted tie, we eliminated that data point from our comparison since there was no way to determine which team the coaches model would predict to win. There were 12 games with the same number of A players on each team so these games were not included in our comparison analysis.

Overall, the coaches' selection technique won 31 games and lost 23 for a winning percentage of 57%. For the same set of games, the NNET model won 39 games and lost 15 for a winning percentage of 72%—much higher than the coaches model. Using all the games, the NNET won 47 games and lost 19 for a winning percentage of 71% (thus showing consistency). These results are shown in Fig. 13.3.

Overall, the NNET model had rankings that resulted in a different team placement for 10 players compared to the coaches model. The difference in player rankings results in the coaches having a much lower winning percentage than our NNET model.

The results and conclusion of this analysis confirm the belief of several sports professionals—namely, that it is extremely difficult to evaluate player performance for players who play more of a supporting and complementary role and perform extremely highly in this role. For example, in soccer, basketball, or hockey, a player may provide the team with great benefit by shutting down the opponent's top player or the player may

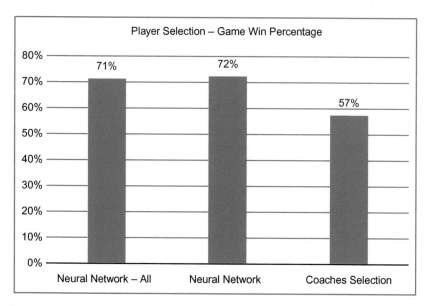

Comparison of NNET Model Performance Compared to Coaches' Selection

Model	Games	Wins	Losses	WinPct
Neural Network – All	66	47	19	71%
Neural Network	54	39	15	72%
Coaches Selection	54	31	23	57%

Figure 13.3 Comparison of NNET Model Performance Compared to Coaches' Selection.

contribute to offense by creating space for their own team's top players to excel. These types of performances do not have any corresponding statistics or metrics, and unless a coach fully observes the player they may not be ranked as highly as their true contribution deserves. This player, however, will surely be highlighted in our NNET approach.

Additionally, there is pretty much a general consensus in the sports industry that we can most certainly reach an agreement for the top 25% of players and for the bottom 25% of players simply by observing game and/or practice performance. However, consensus is rarely found when trying to rank the middle 50% of players into either the second 25% grouping or the third 25% grouping. This is mainly due to not having any appropriate performance statistics that can accurately evaluate the player with respect to team performance. This is where our NNET model will excel.

Finally, while this approach is relatively new to the sports industry, it has proven successful for player rankings and player selections for teams. This technique can be extremely helpful for a national sport system in selecting players for different levels and/or for inclusions or exclusions from a national team. It also provides coaches and national team selection committees with an alternative, unbiased, and completely objective view of player performance and ability, which can complement the current selection process as well as highlight players who may be deserving of further and detailed evaluation.

13.4 ADAPTIVE REGRESSION ANALYSIS

In our sports modeling problems adaptive regression analysis provides a means to revise or update a statistical parameter value (such as the logistic rating parameter) based on a player being added or removed from the team roster for the game. For example, suppose the team's top scorer will not be available for an upcoming game, but the statistical logistic rating parameter was computed based on this player playing in the past. If the player is not going to be available for a game (due to injury or suspension), how can we revise the statistical parameter value to account for the missing player? Similarly, how should the statistical parameter value be revised if a top player will be playing in a game but was not in any of the previous games where the statistical parameter value derived? This would occur if the player is coming off an injury or suspension.

The answer to this question is relatively straightforward, provided that we have team data statistics that have been found to be predictive of team performance. In this case, we can run a regression of the statistical parameter as a function of the predictive team data. Once a relationship is found between the statistical parameter and predictive team data, the statistical parameter value can be revised based on how the team data is expected to change based on the availability or unavailability of a player. However, the effect that the player will have on overall team data and performance is still subjective to some extent, but it will still be able to improve the results.

For example, a regression model of a statistical parameter as a function of team data is as follows:

$$\lambda = b_0 + b_1 \cdot x_1 + \cdots + b_k \cdot x_k + \varepsilon$$

where

λ = statistical parameter value
x_k = team data predictive factor
b_k = sensitivity to team data metric k

Example 13.3 Adaptive Regression Analysis

Let us revisit our NFL modeling results using the logistic ratings parameters and team data statistics (Table 13.2A). If for an upcoming game, we find that a team will be missing a top scorer, we can revise the logistic rating parameter based on its relationship with the team data statistics.

We show results using both team points data and team performance data.

Table 13.2A Adaptive Regression Model

Team	PSPG	PAPG	OFF Y/R	OFF Y/PA	DET Y/R	DET Y/PA	Logistic Rating
Arizona Cardinals	29.4	21.2	4.142	8.075	4.044	6.640	3.368
Atlanta Falcons	21.2	21.6	3.815	7.056	4.038	6.920	1.160
Baltimore Ravens	20.5	25.1	3.859	6.318	3.974	6.857	0.735
Buffalo Bills	23.7	22.4	4.778	7.189	4.391	6.598	1.258
Carolina Panthers	36.9	19.4	4.273	7.336	3.912	5.898	3.937
Chicago Bears	20.9	24.8	3.962	7.004	4.467	7.004	1.468
Cincinnati Bengals	27.2	17.5	3.862	7.573	4.408	6.182	2.777
Cleveland Browns	17.4	27.0	4.024	6.210	4.487	7.851	0.059
Dallas Cowboys	17.2	23.4	4.627	6.563	4.198	7.186	0.226
Denver Broncos	26.4	17.9	4.016	6.460	3.363	5.809	3.447
Detroit Lions	22.4	25.0	3.771	6.665	4.224	7.176	1.734
Green Bay Packers	26.4	20.4	4.339	6.086	4.440	6.771	2.672
Houston Texans	21.2	20.2	3.712	6.078	4.089	6.184	1.424
Indianapolis Colts	20.8	25.5	3.631	5.984	4.319	7.032	1.239
Jacksonville Jaguars	23.5	28.0	4.161	6.768	3.683	7.128	0.172
Kansas City Chiefs	28.9	17.1	4.565	6.978	4.136	5.904	2.674
Miami Dolphins	19.4	24.3	4.349	6.481	4.014	7.380	0.521
Minnesota Vikings	23.4	18.4	4.511	6.431	4.203	6.651	2.654
New England Patriots	31.4	20.1	3.664	6.997	3.889	6.463	2.518
New Orleans Saints	25.4	29.8	3.756	7.451	4.908	8.353	0.955
New York Giants	26.3	27.6	3.993	6.978	4.374	7.497	0.704
New York Jets	24.2	19.6	4.170	6.717	3.579	6.261	1.585
Oakland Raiders	22.4	24.9	3.938	6.401	4.133	6.459	1.520
Philadelphia Eagles	23.6	26.9	3.935	6.560	4.504	6.750	1.002
Pittsburgh Steelers	28.6	19.9	4.532	7.816	3.779	6.805	2.539
San Diego Chargers	20.0	24.9	3.455	6.883	4.808	7.424	0.523
San Francisco 49ers	14.9	24.2	3.959	6.304	4.008	7.612	1.060
Seattle Seahawks	28.6	17.6	4.524	7.634	3.486	6.180	2.711
St. Louis Rams	17.5	20.6	4.559	5.930	4.018	6.809	1.631
Tampa Bay Buccaneers	21.4	26.1	4.756	7.200	3.446	7.098	0.605
Tennessee Titans	18.7	26.4	4.003	6.358	3.890	7.312	−0.448
Washington Redskins	25.4	24.4	3.691	7.361	4.801	7.206	1.360

The regression using team points data is:

$$\lambda = b_0 + b_1 \cdot PSPG + b_2 \cdot PAPG + \varepsilon$$

where

λ = logistic rating parameter
$PSPG$ = team's points scored per game
$PAPG$ = team's points allowed per game
$b_i's$ = regression parameters

Results from the regression are shown in Table 13.2B. Notice the very high goodness of fit with $R^2 = 0.826$. The high R^2 implies that the predictive points factors are also predictive explanatory factors for the logistic ratings parameter. The logistic ratings parameter can then be updated based on the expected effect on scoring of the missing players. For example, if the missing player is expected to result in the team's points scored per game decreasing by $+3$ points and the team's points allowed per game increasing by $+3$ points (since the team will be on offense for less time), the logistic rating parameter can be updated using the sensitivities determined from the regression and the revised predictive input data.

The regression using team performance data is:

$$\lambda = b_0 + b_1 \cdot Off_Y/R + b_2 \cdot Off_Y/PA + b_3 \cdot Def_Y/R + b_4 \cdot Def_Y/PA + \varepsilon$$

where

λ = logistic rating parameter
Off_Y/R = team's yards per rush
Off_Y/PA = team's yards per pass attempt
Def_Y/R = team's defense yards allowed per rush
Def_Y/PA = team's defense yards per pass attempt
b_i's = regression parameters

Results from the regression are shown in Table 13.2C. Notice again the very high goodness of fit with $R^2 = 0.634$. The high R^2 once again shows that the set of predictive factors also serves as explanatory factors for the logistic ratings parameter. The logistic

Table 13.2B Logistic Rating as Functon of Team Points Data

Category	PAPG	PSPG	Const
Est	-0.170	0.118	2.654
Se	0.027	0.020	0.951
t-Stat	-6.258	5.802	2.791
R^2/SE	0.826	0.469	
F/df	68.986	29.000	
SSE	30.320	6.373	

Table 13.2C Logistic Rating as Function of Team Performance Statistics

Category	DET Y/PA	DET Y/R	OFF Y/PA	OFF Y/R	Const
Est	-1.417	0.404	0.583	-0.018	5.707
Se	0.251	0.407	0.231	0.393	3.130
t-Stat	-5.652	0.994	2.521	-0.045	1.824
R^2/SE	0.634	0.706			
F/df	11.677	27.000			
SSE	23.252	13.441			

ratings parameter can then be updated based on the expected effect on scoring of the missing players. For example, if the missing player is expected to result in the team's yards per run decreasing by 0.5 and the team's yards per passing attempt decreasing by 3.5 yards, the logistic rating parameter can be updated using the sensitivities determined from the regression and the revised predictive input data. If the missing player is a defensive player we could make similar adjustments to the defensive team performance statistics.

13.5 CONCLUSION

In this chapter, we introduced three advanced modeling techniques that can be used to help improve sports models. These techniques consisted of PCA, NNETs, and adaptive regression analysis. PCA provides analysts with statistical tools to reduce the underlying data set, which can often be cumbersome, correlated, and filled with data variables that do not lend insight into the problem at hand. Using PCA and principal factors, however, analysts can reduce the original data set into a smaller subset of variables that are uncorrelated and predictive, and can provide insight into the problem that is difficult to ascertain from the original data set. NNETs provide analysts with a modeling technique that mimics the way the human brain solves problems. In addition to providing a different solution technique, NNETs have been found to, at times, provide better evaluation of player performance and ability than via observation of game performance. These techniques are especially useful to assess and rank players in the middle group of ability, where the current performance metrics may not be consistent with the players' true contributions to overall team performance. Adaptive regression analysis provides analysts with means to revise and update statistical parameter values such as the logistic rating parameter based on changing player performance. This is helpful in situations where a player will be in a game but was not in the previous games during which the statistical parameter was calculated, and in situations where a player will not be in a game but did participate in the games where the statistical parameter was calculated. These techniques have been found to improve the models presented in the preceding chapters.

INDEX